T0361896

ROUTLEDGE LIBRARY EDITIONS:
AGRICULTURE

Volume 1

AGRICULTURAL
DEPRESSION IN THE 1920'S

AGRICULTURAL DEPRESSION IN THE 1920'S

Economic Fact or Statistical Artifact?

H. THOMAS JOHNSON

Routledge
Taylor & Francis Group

LONDON AND NEW YORK

First published in 1985 by Garland Publishing, Inc.

This edition first published in 2020
by Routledge
2 Park Square, Milton Park, Abingdon, Oxon OX14 4RN

and by Routledge
52 Vanderbilt Avenue, New York, NY 10017

Routledge is an imprint of the Taylor & Francis Group, an informa business

© 1985 H. Thomas Johnson

British Library Cataloguing in Publication Data
A catalogue record for this book is available from the British Library

ISBN: 978-0-367-24917-5 (Set)
ISBN: 978-0-429-32954-8 (Set) (ebk)
ISBN: 978-0-367-25705-7 (Volume 1) (hbk)
ISBN: 978-0-429-28927-9 (Volume 1) (ebk)

Publisher's Note
The publisher has gone to great lengths to ensure the quality of this reprint but points out that some imperfections in the original copies may be apparent.

Disclaimer
The publisher has made every effort to trace copyright holders and would welcome correspondence from those they have been unable to trace.

AGRICULTURAL
DEPRESSION
IN THE 1920'S ★ Economic Fact or
Statistical Artifact?

H. Thomas Johnson

Garland Publishing, Inc.
New York & London ★ 1985

Library of Congress Cataloging in Publication Data

Johnson, H. Thomas, 1938–
  Agricultural depression in the 1920's.

  (American economic history)
  Thesis (Ph. D.)—University of Wisconsin, 1969.
  Bibliography: p.
  1. Agriculture—Economic aspects—United States—
History—20th century. I. Title. II. Series: American
economic history (New York, N.Y.)
HD1765.J64 1985   338.1'0973   84-48308
ISBN 0-8240-6656-1

All volumes in this series are printed on acid-free,
250-year-life paper.

Printed in the United States of America

Dedicated to

Eric Lampard, Morton Rothstein, and all
other faculty and students who belonged
to a remarkable community of scholars
known as the Graduate Program in Economic
History at the University of Wisconsin

ACKNOWLEDGEMENTS

In the planning, research and writing of this dissertation,
I have incurred an enormous debt to many persons which it is
my pleasure to acknowledge. Eric E. Lampard of the University
of Wisconsin ignited my interest in modern agricultural history
by suggesting many facets of the subject that had been overlooked
by previous scholars. His advice and his careful analysis of
my work have been an invaluable source of encouragement and
guidance. Morton Rothstein of the University of Wisconsin
offerred countless leads and ideas from his vast knowledge of
American agricultural history. For his aid and encouragement
I am deeply grateful. John D. Bowman, also of the University
of Wisconsin, spent many hours helping me understand the virtues,
limitations and methods of the "new" economic history. His
influence appears wherever I use economic ideas to specify and
test historical generalizations; however, he is not responsible
for my often shaky knowledge of economic theory.

For convenient access to plentiful stores of research

material I am indebted to many persons at the Agricultural Library of the University of Wisconsin, State Historical Society of Wisconsin, National Agricultural Library of the U.S. Department of Agriculture, Library of the Board of Governors of the Federal Reserve System, Library of Congress, U.S. National Archives, and the Federal Deposit Insurance Corporation. In this regard, I want especially to thank Miss Helen Finneran of the National Archives and Raymond E. Hengren of the Federal Deposit Insurance Corporation. The FDIC supplied funds and assistance to transfer the worksheets from a WPA study of bank suspensions to Madison, Wisconsin where they are available to qualified research scholars.

Like many agricultural historians before me, I have benefited greatly from generous assistance and advice given by Wayne D. Rasmussen, Chief of the Agricultural History Branch, U.S. Department of Agriculture. In addition, I want to thank Lawrence A. Jones, Ernest H. Wiecking, and William H. Scofield of USDA for giving many hours of their time to help me locate and understand reams of statistical data on farm real estate and farm mortgages.

Many useful insights were gleaned from discussions with

Vernon Carstensen, Fred L. Garlock and Clark Warburton, each
of whom has published numerous works on agricultural and economic
history related to issues in this dissertation. In addition, I
derived much from the comments and criticism of several student
colleagues in the Graduate Program in Economic History at
Wisconsin, especially Richard H. Keehn, Franklin Mendels,
and Joseph A. Swanson.

I owe a very special debt to the Graduate Program in
Economic History at the University of Wisconsin for providing
not only excellent conditions for research, but for the financial
support that made three years of study at Wisconsin possible. The
assistance and encouragement of Rondo E. Cameron, Director of
the GPEH, will always be warmly remembered.

Postscript to the 1984 Garland Reprint edition

An article based on material from various
chapters in this dissertation (esp. chs. 2 and
5) appeared as "Postwar Optimism and the Rural
Financial Crisis of the 1920's" in Explorations
in Economic History, Vol. 11, No. 2 (Winter,
1973-74), pp. 173-192.

CONTENTS

CONTENTS

LIST OF TABLES

LIST OF TABLES

CHAPTER I

AGRICULTURAL DEPRESSION IN THE 1920's:
CONSENSUS AND PARADOX

Evidence that America's farmers did not share in the
prosperity of the "golden twenties" was marshalled in countless books, articles and pamphlets published before 1930.
By the end of the twenties there was widespread agreement in
the United States that the agricultural sector had suffered
economic depression in an era when most sectors of the economy
had experienced unparalleled prosperity. Most contemporary
writers suggested that farmers were deprived of their fair
share of the national income after World War I because of
chronic overproduction of staple farm commodities. Acreage
and equipment that farmers added during the war to supply
overseas markets became redundant when these markets evapor-
ated after 1920; however, the high costs incurred to purchase
these inputs during wartime did not vanish so quickly.
Therefore, it was argued, farmers were compelled to recover
as much of this cost as possible by continuing to use these

inputs in spite of decreasing demand for the output they produced. Furthermore, the increased efficiency of new equipment purchased near the end of the war and on into the twenties complicated efforts to adjust farm output to domestic needs in the postwar decade. Experts writing in the 1920's concluded from these factors that an excessive supply of farm commodities relative to domestic demand caused farm prices to be lower than they should have been to insure adequate returns to farmers.[1]

It is noteworthy that historians have not altered this picture of agricultural depression in the 1920's, even though forty years have passed since it was first painted by contemporary analysts. Louis M. Hacker, writing

---

1. An enormous number of books, pamphlets and articles dealing with the farm crisis were published in the twenties. The following is a sample of the major books, listed by date: U.S. Department of Agriculture, Agriculture Yearbook (Washington, D.C., 1923 and 1924); Edwin G. Nourse, American Agriculture and the European Market (New York, 1924); George F. Warren and Frank A. Pearson, The Agricultural Situation (New York, 1924); Henry C. Wallace, Our Debt and Duty to the Farmer (New York, 1925); The American Academy of Political and Social Sciences, The Annals (Philadelphia, 1925), Vol. CXVII; National Industrial Conference Board, Inc. The Agricultural Problem in the United States and Measures for its Improvement (New York and Washington, D.C., 1927); James E. Boyle, Farm Relief (New York, 1928); John D. Black, Agricultural Reform in the

twenty years later, noted that

> "...while the greater part of the American
> economy recovered by 1922 and was enjoying the
> unprecedented prosperity of the golden nineteen
> twenties, agriculture remained depressed. By
> 1929, the farmer was worse off than he had been
> not only ten years earlier but indeed twenty
> years earlier."[2]

Most experts, even after considering the toll taken by the

economy-wide debacle of the early thirties, hold that economic

conditions for farmers in the "depression decade" were only

an extension of the situation they had faced in the 1920's. For

example, A.B. Genung of the U.S. Department of Agriculture

concluded that in the period after World War I "there ensued one

of the most devastating agricultural depressions of modern

times which lasted, with ups and downs, on into the thirties."[3]

United States (New York, 1929); The American Academy of Political
and Social Sciences, The Annals (Philadelphia, 1929) Vol. CXLII;
Edwin R.A. Seligman, The Economics of Farm Relief (New York,
1929); Clarence A. Wiley, Agriculture and the Business Cycle
Since 1920 (Madison, Wisconsin, 1930).

2. Louis M. Hacker, The Shaping of the American Tradition
(New York, 1947), 1077.

3. A.B. Genung, The Agricultural Depression Following
World War I (Ithaca, New York, 1954), 5.

According to Arthur S. Link, a noted historian of twentieth
century American life, "the most important domestic problem
of the 1920's was the agricultural depression that began in
the summer and fall of 1920 and continued intermittently
until 1935"[4] One might say that the final word on agricultural conditions in the 1920's was spoken by Henry A. Wallace
in his first public address as Secretary of Agriculture in
1933 when he said that "for twelve years American agriculture
has suffered, and suffered cruelly"[5] Agricultural, economic
and political historians ever since have drawn on sources
and interpretations from the 1920's and early 1930's to
depict the decade after World War I as a period of crisis
and depression in American agriculture.

This unanimity of opinion among historians for so
long a period of time is unusual for several reasons. For
one thing, the propensity of historians to rework the
interpretations of their predecessors is so great that it

---

4. Arthur S. Link, American Epoch: A History of the United
States Since the 1890"s (New York, 1955), 263.

5. Henry A. Wallace, Democracy Reborn (New York, 1944), 42.

is surprising to find a major historical issue or event
where the judgement of contemporaries stands unchallenged
for as long as one generation. Revision is an important
feature of historical research and writing. Interpretations
of past events undergo frequent change as historians broaden
their perspectives, uncover new sources or use new methods
to analyse old sources. In recent years, for example,
new insights into many facets of nineteenth century American
economic life have been provided by scholars who make explicit
use of economic theory and statistical inference to develop
new data and reassess existing sources. Equally significant
but often less pretentious examples of historical revision
are also found in the studies of social, intellectual, and
political historians. It is striking that the economic
history of American agriculture in the 1920's has not been
reassessed since 1930.

One could argue, of course, that the conventional
wisdom regarding agricultural depression in the 1920's
required no revision since "all the sources" were covered
and interpreted adequately by contemporary scholars before
1930. This is unlikely, since most historical generalizations

about American agriculture after 1900 rest on a limited

foundation of research.[6] It would not be surprising if accounts

of agricultural distress in the twenties suffered from the

same deficiencies that plague accounts of general agricultural

history for that period. For example, much of the evidence

to support the idea of agricultural depression in the twenties

is drawn from political literature, where farm spokesmen

repeated the eternal cry that all was not well in agriculture.

Historians using this material have done little to segregate

political rhetoric and special interest group pleading from

economic reality. Likewise, economists as well as historians

usually describe and explain farm distress in the 1920's

with ambiguous statistics prepared by the United States

Department of Agriculture. Few scholars have questioned

the value of statistical data that often aggregate conditions

faced by thousands of farmers with widely varying experiences.

It is conceivable that if these sources and data had been

scrutinized and analyzed more severely, the consensus among

economists and historians about economic conditions in

---

6. For example, see remarks by James H. Shideler, "Henry
C. Wallace and Persisting Progressivism: A Comment" _Agricul-
tural History_, Vol. XLI, No. 2, (April, 1967), 121. Also,
see comments in a book review by Morton Rothstein in _The
Economic History Review_, Second Series, Vol. XXI, No. 1
(April, 1968), 204.

American agriculture after World War I might not be as
solid as it is.[7]

Perhaps historians have accepted the traditional
interpretation of agricultural depression in the 1920's
for so long because the basic elements in this interpretation
provided a theoretical foundation for the major farm legislation
enacted after 1933. For example, the framers of the Agricultural Adjustment Act of 1933 laid most of the farm problem
of the early thirties at the door of "overproduction";
however, this idea, as well as the corollary that production
control was the key to recovery, had been well developed by
1929. Two eminent agricultural economists writing in the 1960's

---

7. Note how historical accounts of late nineteenth century
American agriculture have changed in recent years. Historians
once portrayed the farmer of that era in terms drawn from
farm journals, records of farm organizations, political
pronouncements of farm spokesmen, newspapers, and the reports
of government agencies. As a result, few studies examined
the farmer-as-farmer or compared political protest literature
with the underlying reality of the farmers' economic position.
(e.g., see works by John D. Hicks and Solon Buck) Recent
scholarship has modified many of these tendencies. Several
historians, perhaps by viewing sources more objectively
than their predecessors but also by examining sources once overlooked and asking new questions of the material, have shown how
many notions about nineteenth century agriculture were
contradictory and misleading. (e.g., see works by Allan
Bogue, John D. Bowman, Eric Lampard, James Malin, Morton
Rothstein, Mildred Throne, and Harold D. Woodman).

have noted that virtually all of our major farm legislation
since 1933 has aimed at restoring farm income levels by
supporting farm prices and restraining output.[8]

Thus, any historian who questions the traditional
interpretation of agricultural depression in the 1920's
would be attacking the roots of basic farm policies that
have stood for over thirty years. Nevertheless, several economists in recent years have questioned whether surplus output
and price disparity are reasonable explanations for chronic
economic distress in agriculture.[9] Some experts have in fact
questioned whether returns in agriculture have been substantially lower than returns in other sectors since World War I.[10]
Without examining these studies at this point, it is important
to note that reasonable doubt exists about the validity
of many assumptions behind our basic farm policy. Indirectly,

---

8. Earl O. Heady and Luther G. Tweeten, *Resource Demand and Structure of the Agricultural Industry* (Ames, Iowa , 1963), v.

9. *Ibid*.

10. This is discussed below in Chapter II.

this suggests that the traditional interpretation of agricultural conditions in the 1920's, accepted without question
by historians for many years, may be erroneous.

This study represents only the first step toward a
comprehensive economic history of American agriculture in
the 1920's. The main conclusion herein is that the evidence
for economic distress is often misleading and conceals the
actual economic adjustments that occurred in agriculture
after World War I. Further research beyond the limits of the
present work will be necessary, however, before we can produce
a realistic economic history of American agriculture in that
decade. In the meantime, ground must be cleared to make way
for the new edifice. Therefore, this chapter will outline
what I have called the "traditional picture" of agricultural
depression in the 1920's. This outline will define both
the economic conditions that are blamed for the farmers'
adversity, and the statistical variables that are cited as
evidence of agricultural depression. In subsequent chapters
these cause and effect variables will be analyzed in greater

detail. The question throughout the study is whether or not these variables reflected economic depression in the agricultural sector.

References to a "farm problem" in the United States usually point toward the assumption that economic returns in agriculture are much lower than the returns earned in other sectors of the economy. Explanations for this problem do not focus on the inherent abilities of farmers as entrepreneurs, but on the process of industrial development that has characterised U.S. economic growth in the past century. Industrialization itself can imply population expansion, increased urbanization, greater capital intensity and rising

incomes per capita, all of which will increase the demand
for food and fiber. If this demand increases at a faster
rate than the supply of farm output, agricultural terms of
trade, and incomes may rise. The lower the aggregate price
elasticity of demand for farm commodities, the more pronounced
this rise will be. On the other hand, if agricultural producers adopt capital intensive techniques from the industrial
sector, the supply of farm commodities may increase at a
faster rate than demand. In that case, agricultural terms
of trade and incomes may deteriorate. Presumably, the latter
situation has prevailed in American agriculture since
the end of World War I. In other words, the basic problem
of low incomes in farming is attributed to chronic overproduction
of agricultural output.[11]

---

11. This broad definition of the "farm problem" can be
found, among other places, in Earl O. Heady, et. al., Roots
of the Farm Problem (Ames, Iowa, 1965), 3-4; D. Gale Johnson,
"Economics of Agriculture," in Bernard F. Haley, A Survey
of Contemporary Economics, Vol. II (Homewood, Illinois, 1952),
225ff.; and W.J. Anderson, "The Basis of Economic Policy for
Canadian Agriculture," Canadian Journal of Agricultural Economics,
Vol. XI, No. 2, (April, 1963), 19-28.

Historians often use the above supply and demand framework to explain the major trends in American agricultural history since the Civil War. Several periods in the nineteenth century seem to have been marked by widespread economic distress in the agricultural sector. Usually these periods accompanied a wave of frontier expansion, temporarily pushing the nation's agricultural capacity ahead of domestic demand and creating surplus output that was sold in foreign markets at prices that yielded inadequate returns to farmers. By the turn of the century, however, agricultural capacity seemed to be expanding at a rate that was more in step with domestic demand. The agricultural prosperity evident after 1900 is usually attributed to the disappearance of surpluses and greater stability of market demand made possible by rapid domestic urbanization. Historians regard the years from 1900 to 1914 as a halcyon period for the American farmer -- a golden age which in later years became the standard by which to measure agricultural prosperity. Many writers have suggested that prices received by farmers were as close to "ideal" as ever in the five years before the outbreak of World War I.

The war supposedly upset this balanced situation in
two ways. First, the need to supply troops and citizens
of the Allied powers created an enlarged demand for staple
commodities overseas, a demand which was augmented in 1917-
1918 by the United States' own war effort at home and abroad.
This export market enhanced the incomes of American farmers
but it could not be depended on when peace returned and
Europeon production and Atlantic shipping resumed prewar
levels. Although Europe's demand for American farm output
remained at wartime levels for more than eighteen months after
the armistice, reduction of this demand in 1920 is regarded
as a major cause of the sharp drop in farm prices that year.
Secondly, many farmers acquired new acreage and equipment
to supply these enlarged export markets. This seems to have
been especially true for producers of staple commodities
such as wheat and pork. Unfortunately, wartime pressures
inflated the price of land, materials, and equipment as well
as the price of farm commodities. Therefore, many farmers
were presumably stranded in late 1920 with high fixed costs
that take years to recover, while the wartime markets and inflated commodity prices that had induced farmers to incur these
costs vanished in a matter of months. Furthermore, the output
produced with this new land and equipment was now redundant--
a surplus that weakened commodity prices and destroyed any

hope of restoring the 1914 equilibrium.

Other factors presumably exacerbated this disruptive impact of the war; however, most accounts of agricultural distress in the twenties imply that farmers would have been better off if war had not disturbed the balanced situation of 1914. Adding to the allegedly excessive capacity after the war was the growing presence of tractors, fertilizer, and other inputs that allowed farmers to produce more output per man-hour than even before. In addition, there seemed to be a marked decline after 1920 in the United States' per capita consumption of cereal and pork staples that had been in great demand during the war. All together, these forces supposedly pushed the supply and demand pendulum in the opposite direction from where it was moving before the war. Instead of the pre-war problem of producing enough food to meet the needs of a growing industrial society, the problem of the 1920's seems to have become one of overproduction and low incomes in the agricultural sector.[12] The "agricultural depression" of the 1920's appears in most of the literature as the harbinger of the modern "farm problem."

---

12. These contrasts in agricultural conditions before and after World War I are outlined in numerous textbooks and secondary studies such as August C. Bolino, The Development of the American Economy (Columbus, Ohio, 1966), 336-346;

For forty years, historians and economists have concluded that after 1920 American agriculture entered an era of long-run price imbalance and economic maladjustment that was caused by chronic overproduction of farm commodities. Few scholars have specified precisely what the term "overproduction" means in this context, except to suggest that if farmers had produced less than they did in the 1920's better prices and higher incomes would have resulted. Direct evidence of farm surpluses is never found in accounts of agricultural distress published in the twenties.[13] Usually cited are conditions which <u>might have</u> generated surpluses, such as

---

Harold U. Faulkner, <u>American Economic History</u> (New York, 1943), 626-630; Donald L. Kemmerer and C. Clyde Jones, <u>American Economic History</u> (New York, 1959), 586-599; Theodore Saloutos and John D. Hicks, <u>Agricultural Discontent in the Middle West, 1900-1939</u> (Madison, Wisconsin, 1951), 87-110; James H. Shideler, <u>Farm Crisis: 1919-1923</u> (Berkeley and Los Angeles, 1957), 280-295; George Soule, <u>Prosperity Decade, From War to Depression: 1917-1929</u> (New York, 1947), 229-251; and Chester W. Wright, <u>Economic History of the United States</u> (New York, 1949), 520-521.

13. The Bureau of Agricultural Economics (USDA) often received letters in the 1920's from congressmen and others requesting information on the magnitude of commodity surpluses. The BAE always replied with published reports on output, imports and exports and told the correspondent to make his own estimate of the surplus. See letters filed in U.S. National Archives, Files of the BAE, Box 225, "Surpluses."

acreage expansion during the war. Later writers have accepted
without question the overproduction thesis that was embodied
in studies of postwar agricultural distress published before
the end of the twenties. Historians have meanwhile been pre-
occupied with the political figures and legislative battles
that surrounded agricultural problems in the 1920's and less
concerned with the evidence and conclusions put forth in contemporary studies of farm distress. The common starting
point for most legislative programs aimed at farm relief
in the twenties was "overproduction," no matter how imprecisely
the concept may have been defined or how weak the evidence
to support it may have been.

A notable example of this was George N. Peek's famous
plan to insure "equality for agriculture." As embodied in
the McNary-Haugen legislation, this scheme proposed export-
dumping as the remedy for overproduction. Although Peek and
his supporters regarded overproduction as the source of agricultural depression in the twenties, they never advocated
long-run production control as a solution. Instead, they

looked on surpluses as a temporary phenomenon and argued that
reduction of the nation's farm capacity in periods of surplus
could lead to shortages of output in later periods of adverse
weather and crop failure. For Peek, the problem was to alleviate
the price-depressing effects of surpluses which farmers
seemed to face in the twenties, and still maintain the nation's
farm capacity as a hedge against possible shortages in later
years.[14] Thus, he urged that surpluses, especially of a
basic commodity like wheat, be unloaded abroad so that
domestic farm prices would attain levels high enough to assure
farmers a rate of return that was on a "par" with returns
earned by nonfarmers.[15]

---

14. This is discussed in Alonzo E. Taylor, "The Dispensability
of a Wheat Surplus in the United States," Wheat Studies of
the Food Research Institute, Vol. I, No. 4 (March, 1925), 123.
George Peek became first administrator of the Agricultural
Adjustment Administration on May 15, 1933. He resigned the
post only seven months later in a dispute over production
restriction, a measure he opposed. For detailed information
see Gilbert C. Fite, George N. Peek and the Fight for Farm
Parity (Norman, Oklahoma, 1954), 255-266.

15. An excellent discussion of Peek's plan as well as several
other proposals of the time that espoused support of farm
prices is in Murray R. Benedict, Farm Policies of the United
States: 1790-1950 (New York, 1953), 207ff.

Other writers advocated various forms of production adjustment as a means of coping with the overproduction problem. One example was the U.S. Department of Agriculture's "outlook" service, begun in 1923. This service was based on the belief that overproduction at any point in time existed only in certain commodities. In the early 1920's, these were supposedly the commodities that had been in greatest demand during the war, e.g. wheat, pork, and cotton. The assumption was that if farmers were accurately informed about the "outlook" for supply and demand conditions they would reduce production in commodities where surpluses existed or were anticipated and increase production of other items.[16] Later in the twenties, more ambitious schemes of production control were advocated. Perhaps the most important was the domestic allotment plan developed by W.J. Spillman who once directed the USDA's Office of Farm Management.[17]

---

16. Henry C. and Anne Dewees Taylor, The Story of Agricultural Economics in the United States: 1840-1932 (Ames, Iowa, 1952), 447-479.

17. John D. Black, "Plans for Raising Prices of Farm Products by Government Action," in American Academy of Political and Social Science, The Annals (Philadelphia, March, 1929), Vol. CXLII, 382-383.

Eventually, elements from all of these plans were distilled into the Agricultural Adjustment Act of 1933 where voluntary production control was accepted as the principal means of curbing overproduction and raising farm prices.

The ideas implicit in these legislative programs and political debates of the twenties and early thirties did not begin to arise until the mid-twenties, long after the sharp crisis of 1921. This is especially true of the notion, accepted with little question by scholars ever since, that the farmers' problem after 1920 was long-run in nature and could not be corrected by normal adjustment to prices in free markets. Before 1923 no one suggested that agriculture was suffering from anything but a temporary setback that would vanish as the pace of economic activity in the nation quickened. Farm prices, it was true, had fallen relatively farther than most commodity prices in the collapse of late 1920, but on the other hand they had risen more than most prices during the wartime inflation. Historically, normal production adjustments had always rectified such post-war imbalance in a matter of a year or two. Market equilibrium was always restored in the long-run.

This was the sentiment of the Joint Commission of
Agricultural Inquiry called by Congress during 1921 in response
to a Senate bi-partisan farm bloc's demands that an investigation of the crisis be conducted as a basis for remedial
legialation. In its report the Joint Commission determined
from studies of relative price movements that agriculture
had suffered more than other industries in the crash of 1921,
but concluded that the roots of the crisis lay in a business
cycle downturn which, like all past cycles, would naturally
be followed by recovery.[18] Thus, it proposed remedies
to ease the farmer's plight in the interim and, hopefully,
put the agricultural sector on a firmer footing when prosperity
returned. Following the Joint Commission's recommendations,
Congress passed a large package of legislation in 1922 and
1923 that widened the availability of credit to farmers,
forced stringent controls on marketers and processors of
agricultural commodities, and gave agricultural cooperative
associations exceptions from anti-trust and income tax leg-

---

18. U.S. Congress, Joint Commission of Agricultural Inquiry,
The Agricultural Crisis and Its Causes, Part I of the
Report of the Joint Commission of Agricultural Inquiry (Washington,
D.C., 1921), 11-25.

islation.[19]

In spite of these measures, evidence piling up by
late 1923 suggested that agriculture was not sharing in
the recovery occurring in most other sectors of the economy.
For many observers, the key signal that things were not
well on the farm was a statistic that supposedly measured
changes over time in the relative economic position of the
agricultural sector. This economic barometer, known as the
"parity" ratio, was a terms of trade index constructed by
computing for each year the ratio of an index of farm prices
to an index of non-farm prices. By 1922 and 1923 several
price indices had been constructed that made it possible to
compare the relative trends in agricultural and non-agricultural
commodity prices over long periods of time. Taking the
relationship between the farm and non-farm price indices
that prevailed in the decade before World War I as a norm,
it was clear that the relative shifts after 1920 were not
favorable to agriculture.[20]

---

19. Murray Benedict, _Farm Policies of the United States_, chapter 9.

20. John D. Black, _Parity, Parity, Parity_ (Cambridge,
Massachusetts, 1942), chapter 5. James H. Shideler, "The
Development of the Parity Price Formula for Agriculture,
1919-1923," _Agricultural History_, Vol. XXVII, No. 3, (July,
1953), 79-81.

Most users of the parity ratio, including contemporaries in the twenties and later scholars as well, have assumed that price comparisons can gauge how equitably changes in the national product are distributed among the various productive sectors. The conclusion that the agricultural sector was thrust into a position of economic inequality after 1920 was based largely on the failure of the parity ratio to return to the prewar level until after the late thirties. Furthermore, the failure of the parity ratio to regain prewar levels during the twenties was attributed to chronic overproduction of farm commodities after the war.

The parity ratio, however, was not the only statistical sign taken as evidence that economic depression continued in agriculture after the early 1920's. No less alarming were the rising rates of farm mortgage foreclosures, farm bankruptcies, and rural bank failures during the twenties. Also distressing was the continuing fall, albeit at a declining rate, in farmland values in almost every state after 1920. For contemporaries, these trends suggested that farmers were suffering chronic hardship while the rest

of the economy enjoyed unprecedented prosperity. Furthermore,

later historians have noted the similarity of the trends

in these statistics in both the twenties and the early

thirties to support the conclusion that economic conditions

for agriculture were alike in both periods. This viewpoint

was expressed prophetically, so it seems, by the Secretary

of Agriculture Henry C. Wallace in 1924 when he said that

"the depression which began in 1920 was not merely a stretch

of lean years, such as farmers have had to go through before,"[21]

but the beginning of a long-run period of maladjustment.

It is noteworthy that these statistical data are

usually the main evidence to support the notion of agricultural

depression in the twenties. True, many farm organizations

and farm spokesmen who pressed for agricultural relief

measures in the twenties could cite examples of distressed farmers

who fared better during the war than after. But their briefs

and studies relied most heavily on parity, mortgage foreclosure,

bankruptcy, and other such data for evidence of agricul-

---

21. From the annual report of Secretary Wallace in U.S.
Department of Agriculture, Agriculture Yearbook (Washington,
D.C., 1924), 17.

tural depression, These data and the interpretations attached to them by contemporaries have since been used by countless historians, economists, and other scholars who seem
satisfied that they provide sufficient evidence of agricultural
depression in the decade following World War I. Surprisingly,
these data have been subjected to almost no analysis or
criticism in spite of their crucial role. Invariably, the
story they purport to tell is accepted without question as
being obvious.

Unless one accepts certain assumptions from the start,
it is not obvious that these statistical indices, used so
often to describe the farmers' situation in the 1920's,
necessarily reflect economic depression in agriculture after
World War I. For example, the index of the parity ratio
reveals that farm prices in the interwar period never regained
the position relative to nonfarm prices which they held
before 1914. The conclusion generally drawn, therefore, is
that farm incomes were depressed relative to nonfarm incomes
during the entire period. Such a conclusion is not warranted,
however, because the index of the parity ratio is a simple "barter"

terms of trade index that does not consider changes in efficiency of the production inputs whose prices go into the denominator of the ratio.[22] In other words, rising input efficiency can offset the impact of rising input prices on net incomes. Properly considered, efficiency trends might vitiate the usual conclusion drawn by writers who equate a depressed parity ratio in the 1920's with agricultural depression.

Another commonly cited sign of agricultural depression in the twenties is that average farmland values declined throughout the decade -- a trend one might reasonably associate with falling or depressed farm income.[23] In spite of this, the decline in land values after 1920 was perhaps due less to postwar farm income trends than to the phenomenal real

---

22. Inadequacies in the parity ratio as an index of agricultural depression will be discussed in chapter III.

23. The use made of land value, mortgage foreclosure and bank failure statistics in accounts of agricultural distress in the 1920's will be assessed in chapter V.

estate boom in most rural areas in 1919 and 1920. Land
values responded so optimistically to wartime profit levels
that these values _had_ to adjust downward later, unless the
very high rate of increase in farm profits experienced during
the war could have been maintained indefinitely. Under
these conditions, declining farmland values would have prevailed almost regardless of the trend in farm net incomes in
the twenties. That the _rate_ of decline in farm land values
tapered-off as the decade progressed suggests that farm
incomes were in fact rising and might have induced a rise in
values after 1929 had world-wide depression not reversed the
trend in farm incomes. Similarly, the high rate of farm
mortgage foreclosures in the twenties is also related to
the postwar land boom and probably had less to do with income
depression than with adjustments in the land market.

A further statistic that is often associated with
agricultural depression in the 1920's is high incidence
of rural bank failures in the decade. Nevertheless, the
failures were not necessarily related to economic conditions
in farming. It is possible that growing use of the automobile

after World War I led to many bank failures in small towns
as farmers shifted their commercial activities to larger
towns and cities. From this it is clear that <u>rural</u> economic
distress may not always be related to economic conditions
in agriculture itself. Also the fate of many rural banks
was tied to repurcussions in the farm real estate market
in the wake of the postwar land and mortgage boom. Bank
failures in neither case would necessarily reflect depression
in agriculture.

These statistics that are taken so often as unambigious
evidence of agricultural depression in the 1920's will be
studied in detail in subsequent chapters. At stake will
be two questions. First do the trends in these data
justify the usual conclusion that the agricultural sector
of the American economy suffered economic depression during
the twenties less only in degree to that faced in the early
thirties? Second, how valid is the notion that commodity
overproduction was the root cause of agriculture's distress
after World War I? The brief comments above are sufficient
to cast doubt on these assumptions. It will be necessary

in later chapters to go further and see what in fact the trends
in these data reveal. The next chapter will consider one
piece of statistical evidence that is rarely discussed in
accounts of agricultural conditions in the twenties;
namely, farmers' net incomes.

CHAPTER II

THE INCOME SITUATION

The status of farm incomes in the United States
during the 1920's is the keystone on which the interpretation of agricultural distress in that decade must rest.
Most scholars cite falling farm land values, rising farm
mortgage foreclosure and bankruptcy rates, a "weak" parity ratio,
and protests of farm spokesman as evidence of agricultural
depression in the twenties; however, these signs of distress
are always attributed to depressed farm incomes after 1920.
Overproduction of staple commodities and depressed farm
prices presumably drove farmers' incomes to relatively lower
levels than those enjoyed by producers in other sectors of
the economy. Nevertheless, few accounts of agricultural
conditions in the 1920's make explicit reference to the actual
income situation. Therefore, to test the conclusion

that farm incomes lost ground in the decade, the first part
of this chapter will compare the relative trends in farm
and nonfarm incomes in the United States during the 1920's.
Unless these income comparisons reveal what the usual hypoth-
eses about overproduction and depressed farm prices predict,
it is not likely that the traditional account of agricultural
depression in the twenties has much to support it.

Implicit in these comparisons of farm and nonfarm
incomes is a secondary issue -- the farmers' "income problem." Economists often argue that the agricultural sector
suffers growing income disparity as a nation becomes more and
more industrialized. Evidence of this disparity in the
United States has usually been taken from statistics showing
that farmers earn lower incomes per capita than nonfarmers.
Presumably, government efforts to support farm prices at
levels higher than free market equilibria have lessened
this differential, although the question is debateable.
In any event, the major assumption behind America's farm
policy since the early 1930's has been that incomes per
capita in agriculture are lower than in most other sectors.

The nationally aggregated income statistics that invariably cast the farmers' position in an unfavorable light were first generated in the 1920's. Furthermore, the idea that low farm incomes resulted from surplus production and depressed prices took root in the same decade and nourished farm policy arguments for many years thereafter. Thus, the analysis of farm income data which follows will test not only the traditional notion of agricultural depression in the 1920's, but also some basic assumptions behind America's farm policies.

One might expect analyses of "agricultural depression" in the 1920's to focus on trends in farm incomes. Nevertheless, most studies of agricultural conditions conclude only from indirect evidence, such as price ratios and newspaper editorials, that farmers' suffered economic depression after World War I. Few accounts of agricultural conditions during the twenties make explicit reference to the actual income situation.

The reason for this omission is clear; no reliable information on farm net incomes was available before the late 1930's.

The first major series on national income was published
in 1922 by Willford I. King of the National Bureau of Economic
Research.[1] King's data was broken down by industry components,
but did not extend past 1918. King later extended this data
to 1925 in a study published in 1930.[2] Preliminary
versions of these income estimates were available in the
twenties, since one finds them cited occasionally in journal
articles and tracts on the farm problem. For example, many
authors took King's series that showed a declining share of
national income originating in the agricultural sector
over time as evidence of the farmer's deteriorating economic
position.[3]

---

1. Willford I. King, Wealth and Income of the People of the
United States (New York, 1922).

2. Willford I. King, The National Income and Its Purchasing Power
(New York, 1930).

3. For an example see National Industrial Conference Board,
Inc., The Agricultural Problem in the United States (New York,

The U.S. Department of Agriculture published annual data showing net returns to farmers by region beginning with the July, 1925 supplement to _Crops and Markets_. These data, which went back to 1922, were based on information from an annual survey of over 13,000 farmers across the country. With the July, 1926 supplement to _Crops and Markets_, the USDA began publishing annual estimates of gross farm income by state as well as net farm income for the nation for the

---

1926), chapter 2. This dubious use of national income statistics parallels the equally erroneous conclusion many writers draw from census data that show a declining number of farmers in the United States after 1920. Many have assumed that this decline reflected economic weaknesses in the farm sector after World War I. Actually, it is quite normal for the percentage of the labor force in agriculture to decline over time in an industrializing nation. This tendency is dictated by relatively low income elasticities of demand for food and fiber. The long-run effect of rural outmigration in such cases is to equalize marginal opportunities in farm and non-farm enterprises. If people leave agriculture under these conditions, it is appropriate to analyze employment opportunities in the non-farm sectors where they migrate. It is not appropriate to search for "weaknesses" in agriculture in order to explain such migration. Those who remain in agriculture are not necessarily confronted with depression just because the number of farmers is declining. In fact, the number of farmers declined in some regions of the United States for many years before World War I. Westward expansion and opening of new farms on the frontier, however, had always been of sufficient magnitude so that the national figures showed increases in the number of farmers. By 1920 the national totals finally

period after 1919. The sources of these data are not clear,

although some probably came from the National Bureau of

Economic Research.[4] The USDA did not release net income

data for the period before 1919 until the late thirties

when the parity formula in the Agricultural Adjustment Act

was changed from a price to an income basis. At that time

they relased preliminary annual estimates starting in 1910

of farm net incomes for the United States as a whole. These

data show sources of income and production expenses in great

detail.[5] The net income figures (Table II-1) suggested that

---

reflected a trend that has existed in some regions for several
years. In fact the decline in the national figures might
have been noted as early as 1910, except that census reclass-
ifications that year included many persons as farmers who
previously were enumerated elsewhere. See John D. Black,
Agricultural Reform in the United States (New York, 1929),
chapter 1.

4. U.S. Department of Agriculture, Crops and Markets,
Monthly Supplement (July, 1927), 250-255.

5. Bureau of Agricultural Economics, Income Parity for Agriculture, unpublished mimeograph in 6 parts, USDA (Washington,
D.C., 1938-1945).

Table II - 1

AVERAGE PER CAPITA NET INCOME OF THE FARM
POPULATION FROM FARMING
1910 - 1929

1910	$147	1920	$282
1911	121	1921	129
1912	154	1922	158
1913	132	1923	187
1914	145	1924	182
1915	148	1925	243
1916	157	1926	220
1917	278	1927	215
1918	305	1928	224
1919	322	1929	230

Source: U.S. Department of Agriculture, Agriculture Handbook
No. 118, Vol. 3, 79. Total net income of farm operators
includes non-money items, changes in inventories valued
at prices at the farm, and farm wages of farm-resident
workers. Farm population estimated at April 1 of each
year.

farmers' incomes from 1922 to 1929 compared more favorably
with the prewar period than most people had believed; however,
the difference did not seem large enough to refute the picture
of depression in the twenties that had been drawn from other
data sources.

The case for agricultural depression in the twenties
never rested, of course, either on the idea that farm incomes
declined over the decade or that farmers after 1920 earned
less absolutely than they did just before the war. Rather,
the argument accepted by experts and the general public
has been that because overproduction of farm commodities
after the war kept farm prices depressed relative to nonfarm
prices, farm incomes in relation to nonfarm incomes were
lower in the twenties than they were in the prewar period.
This suggests that the most appropriate test of the traditional hypothesis regarding agricultural depression in the
1920's would be to compare the ratio of nonfarm to farm incomes
before and after World War I. Data to make this test have been
available for thirty years; however, scholars to this day persist
in using evidence such as the parity ratio and farm bankruptcy

statistics to confirm the traditional hypothesis.

Table II – 2 shows the nationwide ratio of nonfarm
to farm per capita income annually from 1910 to 1929 as
well as the weighted average ratios for the prewar, war and
postwar periods. The years 1919 and 1920 are included in
the wartime period since European demand conditions still
influenced farm prices and markets in those two years almost
as much as they had during the war itself. The ratios in
the table were calculated from the only available series on
nonfarm and farm per capita incomes, although the data are
not strictly comparable. The reasons for this lack of
comparability will be discussed below; however, the main
effect is to overstate substantially the ratio of nonfarm
to farm per capita income in any given year. Except for one
factor, this lack of comparability does not affect the trend
in the ratio over time. That factor is the inclusion of nonagricultural earnings of farmers in the income totals of
nonfarmers for the years we are concerned with here. It is
safe to assume that the ratio of such earnings to all farm
income grew over the decade of the twenties since automobiles
and improved roads, along with the time saved by increased

Table II - 2

COMPARISON OF FARM AND NONFARM INCOME PER CAPITA,
UNITED STATES, 1910-1929

	Ratio of Nonfarm Income Per Capita to Farm Income Per Capita	Weighted Average Income Per Capita by Period		Ratio of Weighted Average Per Capita Incomes
		Farm	Nonfarm	
1910	3.22			
1911	3.81	$140	$480	3.43
1912	3.09	(1910-1914)		
1913	3.89			
1914	3.28			
1915	3.35			
1916	3.67	$248	$683	2.75
1917	2.32	(1915-1920)		
1918	2.24			
1919	2.42			
1920	3.12			
1921	5.59			
1922	4.54			
1923	4.37	$198	$805	4.07
1924	4.35	(1921-1929)		
1925	3.26			
1926	3.90			
1927	3.81			
1928	3.70			
1929	3.82			

Source: Calculated from data in U.S. Department of Agriculture,
Agriculture Handbook No. 118, Vol. 3, 77-79. Farm income
is the net income of farm operators including nonmoney items
and changes in inventories valued at farm prices, plus farm
wages of farm-resident workers. Farm population was taken
as of April 1 each year. Nonfarm income was drawn from the
NBER national income sources. Included above in nonfarm
income is the income of the nonfarm population from agricultural
sources, such as net rent, mortgage interest, and wages to
nonfarmers. Farm income shown above does not include any
estimate for the amount of income earned by farm people in
nonagricultural work, since estimates are not available
before 1934.

use of labor-saving inputs, widened the horizon of economic
opportunities open to the farmer and his wife and older
children. Therefore, the failure to include nonagricultural
earnings of farmers in the farm income totals imparts a downward
bias to the trend in farm income per capita after World War I.

The first thing to notice in Table II - 2 is that
in spite of this downward bias in the farm net income data,
farm net incomes rose faster than nonfarm incomes per capita
throughout most of the twenties. The _annual_ ratio of nonfarm
to farm per capita net incomes fell in all but two of the
eight years after 1921. Therefore, after the year of adjustment
which followed the postwar boom, farm net incomes failed
to rise faster than nonfarm incomes in only two of the remain-
ing years in the decade. This trend, due in part to migration
out of agriculture, would be even more pronounced than shown
here if farm incomes included the nonagricultural earnings
of farmers. In any event, this trend does not suggest
"depression" in the agricultural sector.

There is, nevertheless, strong evidence in Table II -

2 that traditional accounts of agricultural depression in
the twenties may be valid. Note that the overall average
ratio of nonfarm to farm per capita income in the postwar
years was higher than in the prewar period. Taking nonfarm
income per capita as it actually was in the two periods,
for the ratio of nonfarm to farm income to have been as low
in the twenties as it was in the prewar period (i.e.
for it to have been 3.43 instead of 4.07), average farm income
per capita should have been $235 instead of $198, a difference
of $37 per capita. On the face of it, $37 per capita is not
a large amount, but the difference does exist and calls
for further analysis. The most likely way to explain this
$37 "gap" is to review the gross income and production
expense components of farm net income before and after the
war.

Traditional accounts of agricultural distress in the
twenties would attribute this relative drop in farm income
per capita to overproduction and price disparity. This
implies that average farm income per capita was $37 lower than
it would have been in the twenties because farm commodity

prices, and therefore only gross income, were depressed by
surpluses. Overproduction, as that concept is generally used,
would not necessarily have affected unit production costs.
Therefore, if the income differential of $37 per capita were
due only to factors that depressed gross farm income (receipts),
the ratio of net farm income to gross farm income would have
been lower after the war than it was before the war. Moreover,
the lower net income ratio after the war would have resulted
from a proportionately higher ratio of production expense
to gross income for each item of production expense, unless
there had been some change in the type and productivity of
farm production inputs between the two periods.

The USDA's annual income series that begins in 1910
gives a detailed breakdown of nationwide farm income and
expense figures. Using data from this series, Table II-
3 compares the percentages of total gross farm income that
were absorbed by each of several expense categories in the
prewar years with the percentages absorbed in the 1920's.
It is evident from this table that the ratio of net to gross
farm income in the twenties (42.33) was considerably less
than before the war (50.96), as the "overproduction" thesis
would predict. Within the production expense categories, however,
the percentages of gross income absorbed after the war were not

Table II - 3

COMPOSITION OF FARM NET INCOME BEFORE AND
AFTER WORLD WAR I

	Average Percent 1910–1914		Average Percent 1921–1929	
Total gross farm income		100.00		100.00
Total production expenses		49.03		57.67
Current operating expenses		32.50		36.96
Feed	5.21		7.06	
Livestock	2.76		3.07	
Seed	.83		1.01	
Fertilizer and lime	2.20		2.20	
Repair and operation of buildings	2.61		2.17	
Repair and operation of motor vehicles and machinery	.93		3.47	
Hired labor	10.12		9.98	
Miscellaneous	7.85		7.99	
Depreciation		5.93		7.37
Buildings	3.14		3.32	
Motor Vehicles	.44		1.95	
Other machinery and equipment	2.35		2.10	
Mortgage interest		3.24		4.98
Property taxes		2.98		4.81
Net rent to nonfarm landlords		4.37		3.55
Net farm income		50.96		42.33

Source: Calculated from data in U.S. Department of Agriculture, Agriculture
Handbook No. 118, Vol. 3, 36–38, 41–42, and 44.

uniformly higher than before the war, which suggests that something other than overproduction and depressed commodity prices caused the drop in the net income percentage after the war. It is noteworthy that most of the drop in the net income percentage after the war was due to the higher percentage of gross income absorbed by motor vehicle and machinery expenses (repair, operation, and depreciation) as well as mortgage interest and property taxes. These changes can hardly be attributed to "overproduction" and low commodity prices.

The trend in motor vehicle and machinery expenses can be explained largely by the shift from farm produced inputs to nonfarm produced inputs that accelerated in the late war and early postwar years.[6] For example, production expenses increased when internal combustion power was substituted for farm produced animal power. Theoretically, this extra expense was offset by higher gross income when farmers who owned tractors marketed the crops that previously were fed to work animals. But the ratio of net to gross income would still have fallen, even if the increase in production expenses had been matched by an equal increase in gross incomes.[7] In practice, the increment to gross income from the sale of inputs

6. This is discussed below in chapter VI, part B.

7. Clearly, the value of a fraction falls if the numerator remains unchanged while the denominator is increased.

that were released by the purchase of tractors and gasoline
was usually less than the increment to production expenses.

The inevitable drop in the _ratio_ of net to gross farm
income would not, however, have caused a drop in the _amount_
of net income and, therefore, net income per capita, if the
new inputs were capable of producing more output than the inputs
they replaced. The increasing aggregate productivity of inputs
used in American agriculture in the 1920's as well as the rise
in net farm income per capita in the nation indicate that
this was happening.[8] Nevertheless, the higher ratio of nonfarm
income to farm income per capita in the United States in the
postwar years suggests that some inputs added by farmers
during and after the war were not sufficiently productive.
It is noteworthy that much of the added share of gross farm
income absorbed in the twenties by mortgage interest and
property taxes did not lead to additional net income.

The high level of mortgage interest in the twenties,
especially the early twenties, is related directly to an enormous
increase in the amount of farm mortgage debt between 1918

---

8. Productivity data are discussed below in chapter III.

and early 1920.[9] An increase in debt expense is not necessarily
unproductive if the debt proceeds are used to acquire more
productive inputs or to expand the size of operations in order
to utilize existing inputs more effectively. With most of the
mortgage debt added between 1918 and 1920, however, this was
not the case. The increase in farm mortgage debt was more
likely connected with a boom in the farm real estate market
in which many farmers put up their farms as collateral for
loans that were used in speculative land purchases or to
acquire consumer goods. The boom was presumably based on the
anticipation that farm prices and incomes would remain at
high wartime levels long after the war. When this mirage
faded in late 1920, farm land values dropped sharply. Much
of the debt incurred while the land market was booming remained
on the books throughout the twenties, and the interest paid
was a dead-weight burden for many farmers.

To a lesser degree the rise in real estate taxes was
also for unproductive reasons. There is no doubt that the level
of local governmental services provided in many rural districts

9. The comments on mortgage debt, land speculation, and real
estate taxes in the following paragraphs are supported by
evidence in chapter V.

rose significantly in the twenties, as any series of data on miles of improved roads and numbers of consolidated schools added during the decade shows. Also, these services were in most cases economically beneficial to farmers; nevertheless, to the extent that many of these improvements were contracted for, and debt-funded, in the inflationary period of 1918 to 1920, their real costs may have been excessive in relation to subsequent benefits. Implicit in this is the dead-weight element in the higher level of real estate taxes in the twenties.

Assuming that all of the increase in the percentage of gross income going to mortgage interest and real estate taxes after 1921 was unproductive, an upper-bound estimate can be made of the extent to which these two factors alone caused the ratio of nonfarm to farm income to be higher after the war. Mortgage interest and real estate taxes combined absorbed about 3.57 per cent more gross income in the twenties than before the war, or a total of about $448.5 million. This amounted to about $14.50 per capita, since there was an average of 31 million farm people in the United States in the 1920's, or slightly more than 39 percent of

the $37 per capita depression in farm net incomes. Thus,
as much as two-fifths of the "gap" by which the ratio of nonfarm
to farm incomes was higher in the twenties than before
the war can be traced to two items of expense that affected
farm net incomes independently of any trends in farm prices
and production.

Even if all of the added burden of mortgage interest
and property taxes could be shown to have been for productive
purposes, it does not seem that an increase in the ratio of
nonfarm to farm net incomes that was due to a relative decline
in farm net incomes of $37 per capita could justify the concern
scholars have shown for the idea of agricultural "depression"
in the 1920's. Much of this $37 gap would probably vanish if
nonagricultural earnings of farmers were properly accounted
for in these data. Nonagricultural earnings were equal in
amount to about one-half of the net income of farmers from
farming in the late 1930's when the USDA first began to estimate
these figures.[10] On that basis, farmers' earnings from

---

10. U.S. Department of Agriculture, Major Statistical Series
of the USDA, Agriculture Handbook No. 118 (10 vols., Washington,
D.C.), Vol. 3 (December, 1957), 45.

nonfarm employment would have averaged about $100 per capita in the 1920's. Assuming as we have that the relative importance to farmers of these earnings grew steadily in the twenties, it is conceivable that the $37 "gap" we have considered here would vanish if nonfarm earnings of farmers were taken into account.

In summary, then, the notion of agricultural depression in the 1920's is not substantiated by available data on farm net income. On one hand, net farm income per capita rose at a faster rate than nonfarm income per capita during most of the decade. On the other hand, the apparent postwar decline in the ratio of farm to nonfarm net incomes had little to do with either price disparity or overproduction. Instead, the decline can be attributed in large part to the after-effects of a land and mortgage boom that occured between 1918 and 1920, and the growing importance of nonfarm produced inputs after World War I.

There is no doubt, however, that the large difference between farm and nonfarm earnings revealed in these statistics is startling, whether the differential widened after World War I or not. Historians and economists have rarely challenged

the picture of farm income disparity shown by data like these
that first became available in the 1920's.[11] Likewise, many
scholars still hold that this disparity stemmed from surplus
production and depressed farm prices -- the crux of the "parity"
thesis used to explain agricultural depression after World
War I. Although we have just seen that farm net income statistics
offer little, if any, support for the traditional notion of
agricultural depression in the 1920's, it can still be argued
that the low absolute level of farm earnings in the twenties
(and later) was exacerbated by overproduction of farm commodities.
In fact, many farm relief proposals in the 1920's as well as
most national farm legislation after 1930 worked on the assumption that production control and price supports could rectify
the problem of farm income disparity. To what extent do the
data shown here justify these conclusions?

It is noteworthy that at least one economist in the
1920's suggested that the magnitude of the difference between
farm and nonfarm per capita earnings was exaggerated by

---

11. Data comparing farm and nonfarm per capita incomes were
first published in Willford I. King, Wealth and Income.

data aggregated at the national level. In a study published
in 1929, John D. Black had declared: "If our population
of farm people continued on farms with average real incomes
as low as they are made to appear by the data published by the
National Bureau of Economic Research, the Industrial Conference Board, and others, most of them would properly
be called 'boobs' and 'ignoramuses.' And they are not that."[12]
Later, Black criticized the USDA's income statistics used
above in Table II - 2. Noting that these statistics showed
farmers earning less than one-third as much as nonfarmers
over most of the period from 1910 to 1940, he asked if
such figures could "...possibly tell the whole truth?"[13]
He pointed out several reasons for the exaggereted differential shown in these data, such as the failure to include
in the earnings of farmers any amounts earned in nonfarming
occupations, the failure to account for the relatively

---

12. J.D. Black, _Agricultural Reform_, 30.

13. John D. Black, _Parity, Parity, Parity_ (Cambridge,
Massachusetts, 1942), 111.

greater buying power of farm dollars (a difference less
important today than in the 1920's), and the fact that farm
families, having more children on the average than urban
families, will necessarily have lower per capita incomes,
offset somewhat by lower per capita family maintanance costs.
Allowing for these factors, he estimated that in the 1937-
1940 period real per capita farm income was probably $450
or $500 instead of $217 as reported by the USDA, and that real
per capita nonfarm income was about $635 instead of the reported
$652.[14]

On an earlier occassion, Black has noted that data
on average income per capita in the United States nationwide
would always show farmers at a disadvantage because more than
50 per cent of the nation's farmers were in fourteen southern
states where, for historical and institutional reasons,
farm earnings were desperately low. If the North and the
South were considered separately, farm and nonfarm incomes
per capita within each of these regions would probably not
differ by as much; however, in the South where all income

---

14. Ibid., 113.

levels are low compared with the North, most of the population
is classified as agriultural. Thus, when these two regions
of the nation are added together, per capita nonfarm income
is weighted heavily by high industrial earnings in the North
while per capita farm income is weighted heavily by low farm
earnings in the South.[15] After examining the use some of
his colleagues were making of this type of data, Black
in 1923 asked whether agriculture in the United States
had to "...lift the whole South on its shoulders..."[16]

If Black's analysis were carried a step further, one
might consider the relative position of per capita farm and
nonfarm earnings at the state level rather than the national

---

15. John D. Black, "Income of Farmers-Discussion," _American
Economic Review_, Supplement, Vol. XIII, No. 1 (March, 1923), 182-
183. Also, see J.D. Balck, _Parity, Parity, Parity_, 114-115.

16. J.D. Black, "Income of Farmers-Discussion," 183.

or regional level. Recent work by Richard Easterlin allows

us to compare average farm and nonfarm incomes per <u>worker</u>

for the period 1919-1921 in each of the forty-eight states.

Easterlin's data show that among the thirty-four states of

the Non-South, service incomes per worker in agriculture were

<u>higher</u> than in nonagricultural employment in ten states and

less than fifteen percent below the nonagricultural level

in five other states.[17] In the light of this evidence,

coupled with John D. Black's criticisms, it is difficult

to assign any meaning to the usual national statistics that econ-

omists and historians use to define the farmers' economic

position in the 1920's. It is especially difficult to accept

the view adopted during the twenties and accepted widely

thereafter that America's farmers suffered relative income

disparity because of chronic overproduction and depressed

farm prices. On closer analysis, such income disparity

seems to have existed largely in the South; however, most

---

17. Ricahrd A. Easterlin, "State Income Estimates," in Simon
Kuznets and Dorothy Swaine Thomas, <u>Population Redistribution
and Economic Growth: United States, 1870-1950</u> (2 vols.,
Philadelphia, 1957), Vol. I, 755-756. Easterlin's data are
drawn from Maurice Leven, <u>Income in the Various States: Its
Sources and Distribution 1919, 1920, and 1921</u> (New York, 1925).

of the literature on agricultural distress after World War I
points to the Midwest and West as the major trouble spots.
It would appear that grievances in the latter reions may have
stemmed from other problems than income depression or disparity.
Nevertheless, economists and historians have accepted not
only the evidence on farm income disparity, but also the notion
espoused by America's farm policy makers since the early 1930's
that price and production controls can rectify the problem.[18]

By accepting the conventional wisdom about overproduction,
depressed prices and income disparity, scholars have overlooked
other possible explanations for agricultural distress in the
1920's. This becomes clear when one asks whether there was
no farm "problem" in the states or regions where farm incomes
were relatively much higher than the nationwide statistics
suggest they were. The answer to this depends on the distribution of farm income. If the distribution were very
unequal, high _average_ income per worker may conceal the

---

18. E.O. Heady and L.G. Tweeten, Resource Demand and Structure
of the Agricultural Industry, v.

fact that a few farmers earned very high incomes while a
great many farm operators earned only small returns. National
data on income distribution, available only for recent
years, reveal that to some extent this situation does exist
in the United States. In the years from 1947 to 1960, for
example, 20 per cent of the rural farm families in the
United States received from 47 to 54 per cent of the total
money income earned by the nation's farmers, while 20 per cent
of the urban families received from 40 to 41 per cent of
the money income earned in urban areas.[19] Distribution
data are not broken down by place of residence before 1947,
but national data back as far as 1913 suggest that the above
inequality in the distribution of farm and urban incomes
was about the same in the 1930's and somewhat greater in the
1920's.[20] Thus, there is little doubt that a substantial
number of farmers (most of them in the South) were earning

---

19. Herman P. Miller, Income Distribution in the United
States, U.S. Department of Commerce, Bureau of the Census
(Washington, D.C., 1966), 23.

20. Ibid., 19.

substandard incomes in the 1920's regardless of what the data on average per capita farm incomes showed. It is also reasonable to suggest that the "problem" implied by the existence of these farmers was not an income problem that could be solved by raising the level of farm prices in relation to nonfarm prices. These farmers were among the 50 per cent of American farmers who produced about 10 per cent of all farm crops marketed in the nation in 1929.[21] Price levels obviously had little significance for them.

Since World War II, several economists have focused attention on the problems faced by farmers who are at the short-end of the income distribution spectrum. Modern research suggests that these farmers suffer because of two economic conditions that have existed for several generations. First, the low income elasticity for food at the farm level means that as a nation's real per capita income rises, a smaller and smaller percentage of national income will be spent on

---

21. O.E. Baker, A Graphic Summary of the Number, Size, and Type of Farm, and Value of Products, USDA, Misc. Publication No. 266 (Washington, D.C., October, 1937), 68.

farm products. Secondly, there has been a secular rise in the productivity of inputs in agriculture and, most important, this growth in productivity has been very labor-saving in nature. Thus, fewer workers have been needed over time to produce the volume of agricultural produce required in most industrialized economies.[22] If enough people leave farming, under these conditions the average per capita real income of the farm population need not be lower than average per capita income of the nonfarm population. The crux of the problem, though, is that population, for social, institutional and economic reasons, does not transfer out of agriculture at a fast enough rate to keep average farm incomes on a par with average nonfarm incomes.[23] In the United States, this resource

---

22. Theodore Schultz, _Agriculture in an Unstable Economy_ (New York, 1945), 49 and 81-84.

23. Wyn F. Owen, "The Double Developmental Squeeze on Agriculture," _American Economic Review_, Vol. LVI, No. 1 (March, 1966). Stephen L. McDonald, "Farm Outmigration as an Integrative Adjustment to Economic Change," _Social Forces_, Vol, 34, No. 2 (December, 1955).

transfer problem has been most pronounced in the South,
although it exists to some degree in all farm areas of the
nation.

Thus, closer analysis of the farm income statistics
in Table II - 2 has uncovered several flaws in the conclusions
drawn by traditional users of these data. On the one hand,
the data offer doubtful evidence of agricultural depression in
the 1920's. On the other hand, it seems unlikely that the
disparity shown for farm incomes can be attributed to surpluses
and depressed farm prices. In fact, the disparity shown
by the nationwide statistics seems to exist to a much greater
extent in the South than in the North. Furthermore, relatively
low farm incomes, where they do exist, have been identified in
recent research with resource immobility more so than with
farm price and production maladjustment.[24] It is unfortunate
that contemporary economists and farm spokesmen concentrated
on price and production variables rather than resource input

---

24. An excellent synthesis of this modern view of the "farm
problem" is in Earl O. Heady, Roots of the Farm Problem (Ames,
Iowa, 1965).

and productivity variables in their analyses of farm distress

published in the 1920's; however, it is not legitimate for

a historian to criticize the past.[25] It is legitimate to

criticize historians and economists who later accepted these

contemporary analyses of agricultural depression in the

twenties without regard to flaws that had been uncovered

by subsequent research.[26]

It seems clear that this failure, especially of historians,

to question the data and interpretations of writers whose

works were published in the twenties and early thirties has

precluded until now any meaningful analysis of agricultural

---

25. Nevertheless, one Soviet economist criticized American agricultural economists Henry C. Taylor and George F. Warren for not giving sufficient attention in their writings to the cost-saving potential inherent in farm mechanization. See G.A. Studensky, "The Agricultural Depression and the Technical Revolution in Farming", Journal of Farm Economics, Vol. 12, No. 4 (October, 1930).

26. Some suggestive comments about the failure of agricultural economists "...to challenge basic assumptions and theories" were offered by Vernon Carstensen in "An Historian Looks at the Past Fifty Years of the Agricultural Economics Profession," Journal of Farm Economics, Vol. 42, No. 5 (December, 1960), 994-1006. Carstensen suggests that state and federal funds have rarely gone to "agricultural economists concerned with projects that aimed at trying to understand the workings of economic institutions in relation to farming...Those who aimed at finding out how to work the system were eagerly sought." (998-999) In the light of Carstensen's remarks, it is noteworthy that historians (other than Carstensen himself) merely repeat what economists say, rarely questioning their technical arguments and conclusions.

conditions in the United States in the 1920's. Economic
data aggregated at the national level invariably mask more
than they reveal. Continued use of national, regional, or
even state data on farm incomes, for instance, will not enhance our understanding of farm distress in the decade after
World War I. Wherever possible, future research should focus
on sample statistics at the firm level on a commodity by commodity
basis. The importance of this will be revealed in subsequent
chapters where conditions faced by wheat farmers in the
United States during the 1920's will be discussed. In the next
chapter, however, our attention will turn to another nationwide
statistic so often used to depict agricultural depression in the
1920's -- the "parity" ratio.

CHAPTER III

WHAT ABOUT PARITY?

Orthodox accounts of agricultural depression in the
1920's rely on the "parity" ratio as an important measure
of the farmers' economic welfare. In fact, the parity
concept of agricultural welfare first became prominent in
writings published during the twenties, and scholars thereafter
have used parity statistics in much the same way as they were
used over forty years ago.[1] The parity ratio is a simple
terms of trade figure in which an index of prices that

---

1. The best discussion of the history of the parity concept
is in J.D. Black, _Parity, Parity, Parity,_ chapter 5. Useful
insights can also be gleaned from J.H. Shideler, "The Development
of the Parity Price Formula for Agriculture, 1919-1923," _Agri-
cultural History,_ Vol. XXVII, No. 3 (July, 1953).

farmers receive for the commodities they sell is divided
by an index of prices that farmers pay for the inputs and goods
they buy. The parity ratio is constructed from weighted average
annual price data. Until recently, price ratios in the
1910-1914 period served as the base for the parity ratio.
In the 1920's the parity ratio never reached the level of the
prewar base period, although it came close in 1925. (Table
III - 1)

The depressed level of the parity ratio in the twenties
is often regarded as indirect evidence that farm net incomes
were lower in relation to nonfarm incomes than they had been
before the war. It is safe to say that parity data are the
evidence cited most frequently as proof of agricultural depression in the 1920's. Furthermore, the postwar trend in
the parity ratio seems to suggest more than any other evidence
that the farmers' distressed condition was due to overproduction
and depressed commodity prices. The overproduction thesis,
for instance, states that supply increased farther relative
to demand in the agricultural sector than in the nonagricultural
sector. This placed farm commodity prices at a relatively
lower level than nonfarm prices in the postwar decade.

Table III - 1

INDEXES OF PRICES RECEIVED AND PAID BY FARMERS
AND THE PARITY RATIO, 1920-1929

	Index of Prices Received by Farmers	Index of Prices Paid by Farmers	Parity Ratio
1910–1914	100	100	100
1920	211	124	99
1921	124	155	80
1922	131	151	87
1923	142	159	89
1924	143	160	89
1925	156	164	95
1926	145	160	91
1927	140	159	88
1928	148	162	91
1929	148	160	92

Source: U.S. Department of Commerce, Historical Statistics
of the United States: Colonial Times to 1957, 283: K-129,
K-137, and K-138.

Therefore, as the literature of the period suggested, farmers in the twenties were deprived of "fair exchange value" for their commodities and suffered economic depression as a result.[2]

Nevertheless, a direct comparison of farm and nonfarm incomes before and after World War I has cast doubt on these ideas of agricultural depression and overproduction in the 1920's.[3] The traditional link drawn between the parity ratio and agricultural distress in that decade may therefore be spurious.

The main difficulty in using the parity ratio (or any

_____

2. The foremost account of farm distress during the 1920's in these terms was George N. Peek and Hugh S. Johnson, Equality for Agriculture (Moline, Illinois, 1922). Earlier writers had focused on the concept of "purchasing power equity," but their works did not have the effect of Peek and Johnson's later statement of the problem. For example, see J.A. Everitt, The Third Power (Indianapolis, 1903).

3. Pages 36ff.

barter terms of trade index) to measure economic welfare in
agriculture is that prices alone do not determine a farmer's
net income. This holds true over periods as long as several
years even if the parity ratio is depressed only because
the index of farm commodity prices (numerator in the parity
ratio) has fallen. For example, if general overproduction
supposedly exists, additional output will tend to depress
prices. Furthermore, increased production can not only reduce
farm prices but, because of the inelastic price demand for
most agricultural products, will reduce them proportionately
more than the increase in output. As a result, not only
might the index of farm commodity prices fall in such a case,
but the total revenue from these products will fall. In itself,
this does not spell decreased revenue for individual producers
if a sufficient number of farmers leave agriculture (or diversify)
so that fewer producers share the smaller total revenue. Although aggregate net income per capita statistics suggest that
long-run outmigration from agriculture is never sufficient,
these data ignore inequality of income distribution among

farmers. Low per capita farm incomes probably reflect a
resource adjustment problem that plagues the entire economy
and not an agricultural price and income problem.[4] Therefore,
the parity ratio, even when depressed only because of falling
commodity prices, is an inadequate gauge of agricultural
depression.

Most scholars who accept the "parity" notion of
agricultural depression in the 1920's have not assumed,
however, that falling farm prices depressed the parity ratio
in that decade. In fact, the index of farm prices was fairly
stable after 1921. The parity ratio's failure to return to
prewar levels seems to have been due to the much higher level
of the input price index (denominator in the parity ratio)
in the postwar decade. Therefore, if farm incomes were
depressed in the twenties, it was because input prices had
risen farther after the war than output prices.[5] Nevertheless, there are two flaws in this reasoning. First,
we have shown that national income statistics do not support

---

4. Pages 57-58.

5. For an example of this viewpoint see G.F. Warren and F.A..
Pearson, The Agricultural Situation, 86-88.

the claims that farm net income was depressed relatively or
absolutely in the 1920's. Secondly, no matter what the income
data reveal, the assumption that rising input prices in the
twenties depressed not only the parity ratio but farm incomes
and living standards as well, ignores the effect of changes
in input productivity. In other words, it ignores changes in
the efficiency with which production inputs are converted
to output. An increase in the price of an input need not
reduce net income if the input's efficiency increases also.
Since the traditional parity ratio does not account for
changes in resource productivity, its use as a proxy for farm
net income after World War I is severely limited.[6]

The efficiency of most inputs used in agricultural
production in the United States has increased steadily since
1920, and rising productivity is synonymous with falling
real unit cost of production. If new machines, for example,

---

6. A discussion of the effect of productivity change on a
commodity terms of trade index is in Richard H. Keehn, "Agricultural Sector Terms of Trade for Four Midwestern States: 1870-
1900," mimeograph (Madison, Wisconsin, 1966), 19.

can produce more output per hour than machines they replace,
the real unit cost of machines has fallen. All things equal,
then, a rise in the price of machinery need not increase
unit costs and reduce net income if machine productivity
has increased sufficiently to offset the price rise. All
things do not stay equal, however, since the additional output produced by more productive inputs can increase commodity
supply and thereby reduce the market price. In fact, most
innovations that have increased American farm productivity
have also generated increased commodity output.[7] Nevertheless,
as we have noted, the downward pressure placed on commodity
prices and total revenue by increased output need not affect
the revenue of individual farmers - especially those who cut
costs by adopting improved techniques. In any case, if the
aggregate productivity of American farm inputs rose during
the 1920's, the price "scissors" implied by the depressed
parity ratio did not necessarily reflect downward pressure
on farm net incomes.

There are many series of data that measure changes in

---

7. John Bernard Sjo, Technology: Its Effect on the Wheat
Industry, unpublished Ph.D. thesis (Michigan State University,
1960), chapter 2.

American agricultural productivity. In general, these time
series on productivity measure the extent to which resources
are economized in the production of some type of farm output.
At a single point in time, resource productivity can be measured by calculating the ratio of output to input for some
small period such as a month or year. Changes in productivity
over time can be measured by comparing this ratio between
two or more time periods. For example, the productivity of
labor in wheat production in the United States as a whole
rose from .27 bushels per man-hour in 1800 to .93 bushels
per man-hour in 1900.[8] When economists use historical data
to measure productivity, as in the above example, they usually
measure the _average_ product per unit of input for all of the
input used in a given time period in the state or nation.
Depending on the availability of data, a productivity measure
can be very specific, bushels of hard winter wheat per man-
hour in western Kansas, or it can be highly aggregated,
bushels of wheat per man-hour in the United States. Moreover, measures of productivity can be either partial or general
in relation to the number of production inputs included.
Bushels per man-hour provides a partial measure of productivity
in that it considers only one of the inputs used in crop

---

8. U.S. Department of Commerce, Historical Statistics of the

production, namely labor. A general productivity measure takes into account all of the inputs used -- capital as well as labor -- which in this case might include such items as land, tractors, fertilizer, seed, and harvesting machines. At the highest level of aggregation, productivity data exist which measure total output from all inputs used for the entire agricultural sector of the United States.

Two _general_ productivity series for United States agriculture have been constructed in recent years, one prepared by John Kendrick for the National Bureau of Economic Research and the other prepared by Ralph Loomis and Glen Barton of the U.S. Department of Agriculture.[9] Both series relate the total value of agricultural output (net of intermediate products produced and utilized on the same farm) to the total value of farm inputs for the United States, except that Kendrick's input data exclude nonfarm produced intermediate inputs.[10]

United States: Colonial Times to 1957 (Washington, D.C., 1960), 281: K-87.

9. John W. Kendrick, _Productivity Trends in the United States_ (Princeton, New Jersey, 1961), Appendix B. Ralph Loomis and Glen T. Barton, _Productivity of Agriculture;United States, 1870-1958_, USDA, _Technical Bulletin No. 1238_ (Washington, D.C., April, 1961), 57-58.

10. Charles O. Meiburg and Karl Brandt, "Agricultural Productivity in the United States:1870-1960," _Food Research Institute Studies_, Vol. III, No. 2 (May, 1962), 76.

This exclusion, dictated by the overall purpose of Kendrick's study, imparts an upward bias to his productivity series since nonfarm intermediate inputs have become an increasingly important share of all farm inputs during the twentieth century.

The use of either of these productivity series involves a number of technical difficulties. One problem is to find a common unit by which to aggregate all inputs. Obviously acres of land, numbers of tractors, and man-hours of labor cannot be combined directly. Customarily, the respective dollar values of inputs are added together; however, the use of dollars as a measuring stick can generate problems when camparisons are made over time, since changes in current dollar values do not always reflect changes in the physical volume of inputs but may simply reflect monetary conditions. To avoid this complication, input quantities can be combined and compared over time using constant prices; however, this strategy in turn generates the familiar index number problem. Nevertheless, constant dollar value aggregates of inputs are used in all time series that measure total input productivity.

An additional problem connected with measures of total input productivity is the difficulty of accounting for _all_ the

inputs used to produce any given commodity or group of commodities
(Kendrick's exclusion of nonfarm intermediate inputs notwith-
standing). When studying the productivity of inputs in crop
production, for instance, it is easy enough to account for
the land, the man-hours of direct labor, and the fixed and
current capital that are used. Usually ignored, because of
lack of available data, are the research costs incurred by
public or other agencies to develop better varieties of seed
and to enlighten farmers on more productive methods of operation.
The omission of such items imparts an upward bias to general
productivity indices. Nevertheless, by keeping these problems
and biases in mind, one can still derive a great deal of useful
information from general productivity data.

Turning to Table III - 2, the USDA and Kendrick
(NBER) productivity indexes are shown at decade intervals
from 1899/1900 to 1929. The negligible change in the USDA series
between the turn of the century and 1929 masks the labor-saving
inherent in the shift to a more capital intensive input structure.
Data on labor productivity will be discussed in the next few
pages, while evidence on the changing input structure of
American agriculture after 1900 will be discussed in chapter VI.

TABLE III - 2

GENERAL FARM INPUT PRODUCTIVITY,

UNITED STATES, 1899-1929

	USDA Index (1947-49=100)	NBER Index (1929=100)
1899		84.9
1900	77	
1909		85.2
1910	74	
1919	73.7	85.1
1929	76.0	97.7

Sources: USDA from Ralph Loomis and Glen T. Barton, Productivity of Agriculture; United States, 1870-1958, 57-58. The numbers for 1919 and 1929 are three-year averages centered on the year shown. Prior to 1919, annual figures are not supplied by USDA.

NBER from John W. Kendrick, Productivity Trends in the United States, 365-366, Table B - II. The above numbers are three-year averages centered on the year shown, which accounts for the "1929" level being different from 100.

Kendrick's data, however, exclude nonfarm intermediate inputs
and thereby reflect some of the gains in labor productivity
resulting from greater capital intensity. Nevertheless,
both series suggest there was retardation of productivity
growth in the first two decades of the century, with the wartime
decade showing an unambiguous decline in agricultural input
efficiency. Moreover, both series reflect an increase in
productivity during the 1920's which marked a reversal from the
trend in the two preceding decades.

These _general_ productivity series reveal changes during
the 1920's which are brought into sharper focus by considering
an important _partial_ productivity series which measures the
trend in agricultural output per man-hour. Since it is a
partial measure of productivity, the output per man-hour
series does not measure the inherent productive capacity of
labor itself. The fact that farm output per man-hour has risen
substantially in the United States for many decades does not
mean that the capacity of the agricultural labor force _alone_
to generate farm output has gone up. Rather, it indicates
that the agricultural labor force _in conjunction with_ a growing
supply of other inputs, especially capital, has been able

to produce a growing volume of output per man-hour.[11] Nevertheless, labor productivity data are often cited in the literature, since the rising productivity of farm labor is an
indirect measure of one necessary condition for rising real
farm incomes. Under competitive conditions the return to each
factor of production equals the value of its marginal product.
Growing farm labor productivity, assuming unchanged factor
and product prices, measures the degree to which labor's
relative share of the total return to all inputs in agriculture
has grown.

There are two series on farm labor productivity that
have been prepared in studies of the National Bureau of
Economic Research, both of which show similar trends over the
period 1900 to 1930. The farm output data in the two series
are similar, although one series relates output to the number
of farmers and adult male laborers in agriculture and the other
relates output to man-hours of labor.[12] From 1900 to 1930,

---

11. In spite of this, one still finds cases where improper
conclusions are drawn from the USDA's figures on the number
of people "supported" by the average farmer in the United
States. Pictograms often show this hardy man-of-the-soil
holding on his shoulders a larger and larger number of nonfarm
citizens each decade. More to the point would be a calculation
showing the number of nonfarm citizens needed to support
one capital-endowed farmer.

12. For output per number of workers see Harold Barger and

labor productivity rises 32 per cent in the former series
and 25 per cent in the latter series. In both cases, though,
the percentage increase in the labor productivity index from
1920 to 1930 is about twice as large as the percentage change
from 1900 to 1920, and is not much less than the change which
occurred in the 1930's. When all farm output is divided into
the two broad classes of crops and livestock (including
livestock products), USDA data available after 1910 shows
that labor productivity rose relatively more in crop than in
livestock production between 1910 and 1929. The livestock
index rose from a level of 94 (1935-39=100) in 1910 to 95
in 1919 and then to 103 in 1929. The index for all crops
at the same three dates rose from 76 to 81 and then to 88.[13]

Thus, general and partial measures of agricultural
productivity both show the decade after 1920, in contrast
to the two previous decades, as a period of rising resource
efficiency. One question ignored up to this point, however,
is why agricultural input productivity was growing more markedly
after 1920. Economists offer many possible reasons for the

Hans H. Landsberg, American Agriculture, 1899-1939: A Study of
Output, Employment and Productivity (New York, 1942), 251. For
output per man-hour see J.W. Kendrick, Productivity Trends,
365-366, Table B-II.

13. Reuben W. Hecht and Glen T. Barton, Gains in Productivity

growth in input productivity that has occurred in American
agriculture, yet little effort has been made to sort them
out and measure their particular significance.

One reason often given for increased resource efficiency
is technological change, or change in the state of the arts
surrounding agricultural production.[14] More precisely,
economists define "pure" technological change as a shift
in a production function with constant returns to scale,
where the relative weight of each input remains constant and
no input gains more or less than any other from the change
taking place ("neutral" change). Favorable technological
change, so defined, implies that more output can be obtained
from the same combination of inputs. In other words, technological change can explain some of the increase in a general
measure of agricultural productivity, In fact, if the inputs
and input prices have not changed over time, and we assume

---

of Farm Labor, USDA, Technical Bulletin No. 1020 (Washington,
D.C., December, 1950), 102, Table 57.

14. This discussion of technological change is based on the
following: R. Loomis and G.T. Barton, Productivity of Agriculture,
18ff.; J.B. Sjo, Technology, chapters 1 and 2; Joseph A. Swanson,
"Economic Growth and the Theory of Agricultural Revolution",
Agricultural Economics Research, USDA, Vol. XVI, No. 2 (April 1964).

constant returns to scale, a general productivity index measures

only technological change. Of the two general productivity

indexes discussed above, the USDA's comes closer to measuring

"pure" technological change since it purports to include all

inputs. It is not strictly a "pure" measure of technological change

however, since the weights of the various inputs have changed

over time and technological change in the real world is never

neutral.

Technological improvement has come to American agricul-

ture through the development of better machinery, hardier and

more productive plant and animal varieties, and artificial

fertilizers, among other things. In addition, improved management

techniques leading to work-saving modifications of farm oper-

ating procedures have also been an important example of technolog-

ical change. All such factors have made possible a remarkable,

if not always steady, growth in the productivity of all inputs

used in agricultural production.[15]

---

15. Lester Lave has used Robert Solow's version of a Cobb-
Douglas production function to estimate the share of productivity
growth in American agriculture from 1870 to 1960 that was due
to technological change. (Lave's data are drawn mainly from

A factor that affects _partial_ (as well as general) indexes of productivity is a change in the relative prices of two or more types of inputs. For example, suppose that the maximum possible level of output is being produced with a least-cost combination of labor and capital inputs whose prices are given. Then assume that the price of capital at some later period has fallen relative to the price of labor. Presumably, farmers desiring the same level of output as before will now substitute capital for labor in their production process. The result, however, is that the same output is produced in the later period with an input combination that utilizes less labor. Therefore, an index of labor productivity will rise in such a case. The rise in labor productivity described here, resulting from changed factor price

---

Alvin Tostlebe, _Capital in Agriculture_.) Following Solow's methods, Lave defines technological change as the portion of the increase in farm output per man-year that is not "explained" (statistically) by the increase in farm capital per man-year. This provides a crude estimate, as Lave admits, and one with an upward bias since his data on capital do not include intermediate inputs. Nevertheless, Lave's estimates give a rough idea of trends in technological change from decade to decade. It is noteworthy that Lave's "technological change index" rises by more than 40 per cent between 1900 and 1930; however, most of that change occurred in the decade of the twenties. See Lester B. Lave, _Technological Change: Its Conception and Measurement_ (Englewood Cliffs, New Jersey, 1966), 48-52, 74-75, Table 7.2.

relationships, has long prevailed in American agriculture, as in many other sectors.

There are many specific sources of productivity change besides the general factors mentioned here. Some more specific factors relevant to the 1920's can be considered if we reflect on the USDA's crop and livestock labor productivity indexes discussed above.[16] For example, the increased labor productivity noted in livestock production can be attributed to such things as better commercial feeds and feed supplements (resulting in more pounds of meat per pound of feed), eradication of certain diseases through immunization (such as the successful attack on hog cholera), and improved management (as in the growing practice of breeding sows to two litters a year instead of one). In dairy production the advent of the milking machine was of some importance; however, its use as late as 1929 was still confined to the limited numbers of rural areas

---

16. An excellent source on factors behind changing agricultural productivity in the United States is John A. Hopkins, Changing Technology and Employment in Agriculture, USDA (Washington, D.C. May, 1941). Most of the comments in this paragraph are drawn from Hopkins's study.

served by electric power.[17]

In crop production, the growth of labor productivity resulted primarily from increased mechanization, especially of seeding and harvesting operations in small grain production. Also, the westward shift of small grain production into areas of the Great Plains where such machinery could be used to greatest advantage was a factor in raising the overall level of labor productivity in crops. The relatively larger gains in labor productivity in crop production before 1929 reflect that the opportunities for labor-saving inherent in mechanization were less available to livestock and dairy producers in that period. In the case of both crop and livestock production, however, it is important to note once again that the labor productivity gains were much larger in the decade of the twenties than from 1900 to 1920.

---

17. By the end of 1924 only 3.2 per cent of all farms in the United States had central-station electric service. There is no data on the number of farms with auxiliary generating equipment. Martin R. Cooper, Glen T. Barton, and Albert P. Brodell, Progress of Farm Mechanization, USDA, Misc. Publication 630 (Washington, D.C., October, 1947), 55.

This by no means exhausts the list of possible explanations for rising input productivity in agriculture during
the 1920's (or any time period); yet, in spite of many generalizations about the sources of improved agricultural productivity,
almost no work has been done to isolate the various factors and
measure their relative importance. Economists have done a
great deal to measure the absolute amount of productivity change,
although much remains to be done to explain the changes that
have been measured.[18] While no attempt has been made here to
close this gap, the problem must be kept in mind when considering the data on productivity.

This sweeping review of data on agricultural resource
productivity has revealed that input efficiency improved
significantly in the decade after World War I, although the
reasons for the improvement have not been specified with any
precision. Nevertheless, evidence of rising farm input produc-

---

18. A recent attempt to study the sources of productivity
change in American agriculture is William Parker and Judith
Klein, "Productivity Growth in Grain Production in the United
States, 1840-1860 and 1900-1919,", in National Bureau of
Economic Research, Output, Employment and Productivity in the
United States after 1800 (Princeton, New Jersey, 1966).
The Agricultural History Research Center in Davis, California
noted the lack of historical study of sources of productivity
change in agriculture and in 1966 issued a bibliography on
technology in agriculture to encourage future research.

tivity casts doubt on the conventional use made of the parity ratio to depict agricultural depression in the 1920's. If prices alone determined net income, then the low level of the parity ratio in the 1920's might be sufficient evidence that farm incomes were depressed in that decade. The rising level of resource productivity in the twenties suggests that for many farmers higher prices for production inputs were offset, partially at least, by increased efficiency. Therefore, the "parity" picture and the net income picture must be considered separately. This brings us back to a question posed earlier; namely, why the significance of improved resource efficiency was ignored not only by most contemporary analysts of farm distress in the twenties, but by later historians as well.

The attitude about the prospects for improved resource efficiency in agriculture that was most common in the 1920's was clearly stated in a public document published early in the decade. The Joint Commission of Agricultural Inquiry, created by Congress in 1921 to investigate and suggest remedies for the economic crisis affecting agriculture, concluded in its final report that "on the whole, it seems probable that

increased production of farm products in this country must be
at the expense of increased costs." The report went on to
state that "there is nothing...which justifies a conclusion
that greatly increased fixed charges resulting from higher
land values, greater indebtedness, larger outlay for equipment
and machinery, fertilizer, and labor, ... can be paid for out
of either increased production per man or per acre. Payment
of these increased costs must come out of larger money returns.."[19]

Few people in the 1920's realized the labor-saving
advantages inherent in many of the new capital inputs that
were becoming available to farmers. Overlooked was the extent
to which these inputs could reduce real unit costs of production.
Most contemporaries, viewing the world in the static terms of
classical economics, saw the income problem in agriculture
as one of output and prices, not one of input productivity
and unit costs.[20] Fluctuations in farm net income were therefore
attributed to changes in the relation between output and input
prices, expressed in terms of the "parity" ratio. Since the

19. U.S. Congress, Joint Commission of Agricultural Inquiry,
The Agricultural Crisis and its Causes, 204.

20. There were frequent references in the twenties to diminishing returns in American agriculture and the need to keep people

productive capacity of inputs was considered fixed, and there was little hope of influencing the prices farmers paid for these inputs, it is not surprising that many farm relief measures emphasized price and/or output manipulation as the principal means of raising farm net incomes.

For the members of the Joint Commission, larger money returns were to be found mainly in "...better organization and more efficient marketing of farm crops."[21] This panacea was urged by many groups in the twenties who felt that farm incomes could be significantly improved only by reducing marketing margins or by distributing sales of crops more evenly over the year, in order to avoid the low prices otherwise received when a large percentage of a crop is disposed of at harvest time.[22] Other groups, less sanguine about

from leaving the farm. For examples, see George F. Warren, "Some After-The-War Problems in Agriculture," Journal of Farm Economics, Vol. I, No. 1 (June, 1919), 14; Editorial noting Secretary of Agriculture Meredith's departure from office, The Sunday Star (Washington, D.C,, February 27, 1921), Part 2, 1. Filed in U.S. National Archives, Files of the Bureau of Agricultural Economics, Box 224, "Agricultural policy: 1917-21"; Henry C. Wallace, Our Debt and Duty, 17.

21. U.S. Congress, Joint Commission of Agricultural Inquiry, The Agricultural Crisis and its Causes, 204.

22. M.R. Benedict, Farm Policies of the United States, 194-198.

the prospects for saving money through increased marketing
efficiency, saw the way to larger money returns for farmers
in various schemes to manipulate the market prices of farm
commodities. Several proposals for price control and price
fixing were put before Congress in the early twenties, the
best known of which was the two-price export dumping scheme
advocated by George Peek and embodied in the several Mc-Nary-
Haugen bills.[23] One leading agricultural economist even
went so far as to encourage farmers to enhance their economic
welfare by not investing in additional land and capital
inputs. According to Henry C. Taylor, continued investment in new production facilities would only drive up input
prices and result in extra output, falling commodity prices
and reduced net incomes. He urged farmers instead to raise
their standards of living by investing in such things as
home improvements and education for their children.[24]

---

23. Ibid., 198 and 207ff.

24. Taylor reiterated this idea on several occasions in the
early twenties. For an example, see H.C. Taylor, "The Farmer's
Economic Problem," in Report of the Kansas State Board of Agri-
culture (Topeka, March, 1920), 34-35. Filed in the Henry C.
Taylor papers (State Historical Society of Wisconsin, Madison),
Box 28.

The failure to consider changing resource productivity
and the undue emphasis on commodity prices and output in
farm relief measures during the twenties is understandable,
considering the limited changes in agricultural technology
that had occurred in the generation before 1920. The lag
in technological change during the period immediately preceding
the age of the tractor had blunted most people's awareness of
the significance of cost-saving innovation in farming.[25]
One historian has noted that even among farm economists of the
time, many "...were reluctant to acknowledge that the tractor
had been invented and there seems to have been a tendency
among some to regard it merely as a different kind of horse –
one that used kerosene instead of oats."[26] Furthermore, few
historians have ever questioned the relevance of this viewpoint
vis-à-vis the twenties. The spectacular growth in agricultural productivity that occurred after the mid-thirties.

---

25. Earle D. Ross, "Retardation in Farm Technology Before the
Power Age," Agricultural History, Vol. 30, No. 1 (January, 1956).

26. Vernon Carstensen, "An Historian Looks at the Past Fifty
Years of the Agricultural Economics Profession," 998.

has caused most scholars to overlook, or at least underrate,
the significance of technological changes in some branches
of farming that did occur in the decade after World War I.
This oversight cannot continue, however, if we are to properly
assess the meaning of agricultural "depression" in the 1920's.

This chapter has shown that failure to account for
productivity change negates much of the significance attached
to the parity ratio as an index of agricultural depression
in the 1920's. In this way, the parity concept has been
criticized on its own terms. It is noteworthy, however, that
in spite of our evidence on productivity change, parity data
even when adjusted for changes in input efficiency would not
give a valid measure of trends in farmers' welfare. The parity
ratio, because it is aggregated at the national level,
must always be an ambiguous statistic . Similarly, this

caveat applies to almost all indexes of farm distress in the
twenties. National data include all farmers as if the demand,
cost, and technology conditions faced by each one were identical.
In spite of this, the agricultural sector is comprised of
many industries facing widely varying conditions. An aggregate
statistic that includes data on all of these industries at
once, whether it be for income, parity, land values, or anything
else, can reveal nothing but confusion.

The problems inherent in using nationally aggregated
farm statistics will become clearer in the next chapter where
conditions faced by wheat producers in the United States
during the 1920's will be discussed.

CHAPTER IV

OVERPRODUCTION AND PRICES:
"THE WHEAT SITUATION" REVISITED

Writers frequently refer to the plight of the wheat

farmer when they describe the type of farm distress in the

1920's that the phrase agricultural "depression" implies.

Data on mortgage foreclosures, farm bankruptcies and price

disparity all suggest that wheat farmers suffered more than

any single commodity group from the postwar agricultural

crisis. Furthermore, the wheat farmers' problem seems to

clinch the argument that farm distress in the twenties

resulted from overexpansion and overproduction. There is no

doubt that wheat acreage and exports rose to unprecedented

levels during the war. It is significant, therefore, that

this contribution to the war effort seemed to make wheat

farmers more vulnerable than most producers to the threat

of postwar overproduction and depression. One economic hist-

orian, writing as recently as 1968, has reiterated this

viewpoint when he asserts that "the immediate origins of the farm problem that overtook the United States after 1920 lay

in the expansion of staple production during World War I

and in the subsequent erosion of the overseas market for surplus

agricultural commodities. The plight of the wheat farmer

was especially acute...."[1]

Data on the wheat situation provided much of the raw

material for contemporary writers who developed the main

elements of our traditional picture of agricultural depression

in the twenties. When historians relate the wheat growers'

problems in their accounts of agricultural depression in the

1920's, they merely repeat what most writers had already

said before 1930. Most research on the twenties has focused

on political spokesman for the farmer and farm legislative cont-

roversies; little has been done to analyze economic conditions

among individual commodity groups in agriculture, including

wheat farmers. If there were any flaws in analyses of the

wheat situation written over forty years ago, then much that

has been said about agricultural depression in the twenties

---

1. William Greenleaf, ed., American Economic Development
Since 1860 (New York, 1968), 309. The quotation is from
Greenleaf's own editorial comments.

must be reconsidered. Accordingly, this chapter will present
price and production data to test some of the conventional
generalizations about overproduction and economic depression
in wheat-growing areas of the United States during the 1920's.
The following analysis of conditions faced by wheat growers
after World War I will bring to a head the material in
preceding chapters where criticism was levied against the
notion of agricultural "depression" in the twenties. This
case study of wheat, supposedly the most "distressed" sector
of American agriculture in the twenties, will help reinforce
our previous criticism of vague concepts like parity and overproduction.

Of all the articles and books that have influenced
public opinion about agricultural depression in the 1920's,
undoubtedly the most important one has been "The Wheat Situation."
This study of economic conditions among America's wheat growers
was published by the U.S. Department of Agriculture in late

1923.[2] Historians often regard "The Wheat Situation" as one
of the earliest analyses to consider the postwar agricultural
crisis as a long-run maladjustment rather than a temporary
cyclical disturbance.[3] At the end of 1922 most experts
still agreed that agriculture's postwar setback, although
more severe than in other sectors of the economy, would
disappear as the general level of economic activity returned
to normal. This optimism waned during 1923, however, as evidence
mounted that farmers were not sharing in the prosperity that
was being enjoyed elsewhere in the economy.[4] Accordingly,
the USDA undertook a study of this persistent farm crisis
since it seemed by mid-1923 that full recovery for farmers
would require more than voluntary adjustments to normal market
forces. "The Wheat Situation" not only suggested that the
economic crisis in agriculture, especially in wheat, might last

2. U.S. Department of Agriculture, "The Wheat Situation,"
Agriculture Yearbook: 1923 (Washington, D.C., 1924), 95-150.
The bulletin was also released as a Yearbook separate in November,
1923.

3. For comprehensive discussions of the origins and effects
of this bulletin see: Henry C. Taylor, A Farm Economist in
Washington, 1919-1925, unpublished manuscript, Henry C. Taylor
papers, State Historical Society of Wisconsin (Madison),
Box 36, chapters 20-24; J.H. Shideler, Farm Crisis: 1919-
1923, 260-268; Donald L. Winters, Henry Cantwell Wallace
and the Farm Crisis of the Early Twenties, unpublished Ph.D.
thesis (University of Wisconsin, 1966), chapter 9.

4. Refer above to pages 19-21.

a long time, but it also endorsed a plan for recovery that involved government action in farm commodity markets.

On the surface at least, the USDA was prompted to write "The Wheat Situation" by signs of continuing economic distress in the wheat-growing regions west of the Mississippi These signs included a rise in farm bankruptcy rates, particularly in the Northern Plains and Mountain states, and a sharp downturn after May, 1923 in the "purchasing power" of a bushel of wheat.[5] One author summarized the main factors that drew public attention to the wheat problem in 1923 by noting the following:[6]

> "The wheat growers were experiencing a third
> successive year of low prices. Returns were
> lower than in any year since 1913, and lower
> relative to the prices of nonfarm commodities
> than at any time since the depression years of
> the 1890's. Wheat acreage was being reduced,
> but production had remained close to the 1920
> level..."

The conclusion reached in "The Wheat Situation" was that "this condition of things has resulted from the decline in

---

5. H.C. and A.D. Taylor, The Story of Agricultural Economics, 591.

6. M.R. Benedict, Farm Policies of the United States, 211.

wheat prices, the relatively high level maintained in the prices
of other commodities and services, and also from the malad-
justments which exist in the wheat industry itself."[7]

The chief maladjustment "in the wheat industry itself"
was viewed as overproduction, caused largely by wartime expansion and the difficulty farmers faced in reducing their
production capacity to peacetime demand. Although "The
Wheat Situation" acknowledged that farmers themselves could
do much to alleviate the problem of surpluses, favorable
results would not come soon.[8] In the meantime, farmers would
continue to bear a costly burden thrust on them largely by
the war. Therefore, the bulletin stated that "...what is
most needed right now is some way to restore the proper
 price ratios..," not just for wheat but for any commodities
suffering postwar production imbalances.[9] The recommendation
in "The Wheat Situation" was to raise prices by dumping surpluses
abroad through a government export-corporation.

7. USDA, "The Wheat Situation," 145.

8. Ibid., 146.

9. Ibid., 149.

One reason "The Wheat Situation" gained prominence
in the twenties, and a reason for its continued importance
to later historians, was its endorsement of the controversial
plan for agricultural recovery that was being championed by
George Peek, formerly an executive to the Moline Plow Company
and Industrial representative to the War Industries Board in
1917 and 1918. Peek argued that the postwar farm crisis
arose because farm prices fell farther than nonfarm prices
after 1920. This led to a condition of price diaparity that
had impaired the farmers' "purchasing power."[10] The controversial aspect of Peek's plan, however, was not that he viewed
the farm crisis in terms of postwar overproduction and price
disparity, but that he proposed government action to raise
farm prices as a solution.

In essence, he proposed a formula to determine fair
"ratio prices" (later referred to as parity prices) for
staple farm commodities and recommended that a government
corporation be established to purchase any amount of these

10. For sources on Peek and his famous "plan" see citations
above on pages 17 and 64. I am not aware that any historian
has ever noted the striking similarity between England's
Agriculture Act of 1920 and the "parity" principle outlined
by Peek in 1922. See Gerald Egerer, "The Political Economy of
British Wheat, 1920-1960," Agricultural History, Vol. XL, No. 4
(October, 1966), 295.

commodities that farmers could not sell domestically at the ratio prices. The corporation would sell these "surpluses" abroad at world market prices and charge its losses pro rata to all producers of the commodities involved. Peek always cited wheat as the commodity that would gain most from such a plan, since wheat prices seemed more depressed than most after the war and the traditional reliance of American wheat farmers on export markets had been rekindled by wartime markets. Therefore, "The Wheat Situation," after analysing the apparent deterioration of economic conditions in wheat-growing areas in 1923, endorsed Peek's export-corporation plan as the most feasible of all the legislative proposals then being offerred as solutions to the crisis.[11] Since Peek's export-corporation plan appeared in all of the McNary-Haugen bills, it became a center of controversy in farm legislative battles during the twenties; however, none of these bills ever became law. Nevertheless, Peek's persistent efforts to sell his idea helped create public acceptance of the "parity" concept of agricultural welfare and led to the adoption of "parity prices"

11. USDA, "The Wheat Situation," 150.

as the major target variable in American farm policies after 1933.[12]

This acceptance of the "parity" concept as a keystone of farm policy in the thirties explains in part the importance historians have ataached to "The Wheat Situation" as an analysis of agricultural depression in the 1920's. The picture of the postwar farm problem viewed as "correct" by most legislators in 1933 had clearly been outlined in "The Wheat Situation" where farm grievances after 1920 were attributed to price disparity and overproduction. Furthermore, the admonishments in that early bulletin that postwar production imbalances would not soon be corrected without government action seemed prophetic, so it appears, in light of renewed concern about wheat surpluses and low prices in the late twenties. Historians therefore accept "The Wheat Situation" as an authoritative and credible analysis of agricultural depression in the 1920's.[13]

Nevertheless, we have noted how modern theories about the "farm problem" suggest that conventional accounts of post-1920 agricultural depression put a mistaken emphasis on prices

12. G.C. Fite, George N. Peek and the Fight for Farm Parity, 243-266 and 303.

13. For example, Wm. Greenleaf in American Economic Development, 309.

and overproduction as causal factors.[14] Thus, the story of
surpluses and price disparity in "The Wheat Situation" may
have been poor ground on which to base so many of our general
ideas about economic distress occurring on the farm, even in
wheat areas, during the twenties. Elaborate economic analysis
is not required, however, to find flaws in the picture that
"The Wheat Situation" presented. For example, crude national
statistics, which are analysed below, seem to contradict the
idea that wheat production was not adjusting adequately by
1923. Further, it is not obvious that the wheat surpluses which
accumulated in the late twenties were a result of long-run
production imbalances induced by wartime expansion. Before
turning to any data on wheat production and prices, however,
we should predict what we expect such data to show under conditions
like those which prevailed in the United States after World War I.

Basic price theory tells us that a firm in a competitive
industry, such as wheat farming, maximizes profits by producing
output at the level where marginal cost equals market price
(marginal revenue).[15] Taking the size of the firm as given in

14. See above on page 58.

15. Except for certain elaborations for which sources will
be given, the following analysis of firm behavior can be
found in standard texts on price theory such as those written

the short-run, the maximum profit level of production will occur
in the range of diminishing returns to the fixed factors of
production. In that range, marginal cost rises as the level
of output increases. A competitive firm by definition cannot
influence market price; it must act as a price-taker. Therefore,
a profit-maximizing competitor, producing at the point where
marginal cost equals market price, will tend to increase
output as market price rises and decrease output as market
price falls. In the early twenties the market price of wheat,
for instance, dropped sharply as total demand for wheat
decreased. We would therefore expect in that period that
rational wheat farmers would have reduced acreage in order
to maximize profits.[16] As a result of all producers doing
this, the wheat industry supply would have decreased, causing
market price to fall less than it would have if only the
original decrease in demand had affected the price equilibrium.

In fact, a reduction in wheat acreage and production
is precisely what we find in the United States in the early
twenties. The average farm price per bushel of wheat in 1921

by H.H. Liebhafsky, George Stigler, Richard Leftwich, and Joe Bain.

16. Changes in production of wheat do not always correlate
exactly with changes in acreage because yields can at times
affect production considerably more than acreage. The best
index of farmers' production intentions is acres planted.

was \$1.10, down more than 50 per cent from the average farm

price of \$2.14 in 1920.[17] Annual average prices of course

hide the fact that wheat prices had begun to fall by August,

1920 -- soon enough to affect winter wheat planting decisions in

that year. Thus, it is significant that acreage planted to

wheat in the United States in 1920 was more than 13 per cent

less than in 1919. The drop in wheat prices that began in

late 1920 had tapered-off significantly by mid-1921, so that

the drop in acreage planted was much less in 1921 than it had

been in 1920. Nevertheless, acreage planted to wheat continued

to decline until 1925. By 1924, however, wheat prices had begun

to rise steadily for the first time since the war.[18] As

one analyst noted at the time, "the harvested acreage of

wheat of 1924 was 28 per cent less than in 1919; the per capita

wheat acreage harvested in 1924 was slightly below that of

the average of the five years before the war. A large part

of the readjustment has been completed."[19]

---

17. Frederick Strauss and Louis H. Bean, Gross Farm Income
and Indices of Farm Production and Prices in the United States,
1869-1937, USDA, Technical Bulletin No. 703 (Washington, D.C.,
December, 1940), 36, Table 13.

18. Monthly wheat price data from U.S. Department of Agriculture,
Agriculture Yearbook: 1923, 624, Table 34. Acreage planted from
U.S. Department of Agriculture, "Wheat: Acreage, Yield and Production
by States, 1866-1943," Statistical Bulletin No. 158 (Washington,
D.C., February, 1955), 2.

19. A.E. Taylor, "The Dispensability of a Wheat Surplus in the

These facts not only run counter to the pessimistic
picture painted in "The Wheat Situation," they also belie
the notion often parrotted by historians and economists who
say that farmers (especially in the early twenties) tend to
maintain or even increase production in the face of falling
market prices.[20] This notion is usually based on the assumption
that farm industries are less concentrated than nonfarm industries and thus find it more difficult to reduce output when
prices fall. Wheras the few firms in a concentrated industry
can lay-off workers and curtail production when prices fall,
it is assumed that individual farmers, unable to "fire"
themselves and their family labor force, continue producing
output that only gluts the market and worsens their condition.
In reality, however, farm operators are no less rational than
nonfarm producers and rational profit-maximizing behavior
leads to output reduction when prices fall no matter what
type of industry a firm operates in. A producer who responds
to falling prices by raising production is behaving rationally
only if he faces the unlikely situation that whenever prices
fall his marginal costs fall with rising output.

United States," 135.

20. For example, see Earle D. Ross, "Agriculture in an Industrial
Economy," in Harold F. Williamson, ed., The Growth of the American
Economy (Englewood Cliffs, New Jersey, 1951), 687.

TABLE IV - 1

SHORT-RUN PROFIT MAXIMIZATION FOR

A COMPETITIVE FIRM

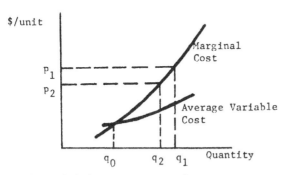

q_1: Profit-maximizing output at price p_1

q_2: Profit-maximizing output at lower price p_2

q_0: Minimum output level in the short-run where marginal cost equals average variable cost

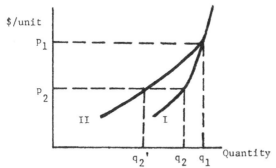

I: Very inelastic marginal cost function, no production alternatives, output falls from q_1 to q_2 when price falls from p_1 to p_2

II: Less inelastic marginal cost function, production alternatives exist, output falls from q_1 to q_2' when price falls from p_1 to p_2

It is true in some cases, of course, that farm output does not fall very much even when prices fall a great deal. This can be explained, however, by the highly inelastic marginal cost function and the high percentage of fixed costs that many farmers face in the short-run.[21] An inelastic marginal cost situation can arise where there are few alternatives to the production of one commodity on a farm, as in the case of specialized wheat farms on the Great Plains. There, farmers must continue producing wheat even when the market price falls considerably, or else stop production. The decision to stop production in the short-run depends on whether prices have fallen to the point where average variable costs are no longer being covered. In many types of farm operations, including specialized wheat farms, short-run variable costs are a small percentage of the total cost of production so that market price must fall very much -- even more than wheat prices did in 1920/1921 -- before production stops altogether. Therefore, inelastic marginal cost functions and high fixed cost ratios will lead to situations where sharp decreases in price will not necessarily be accompanied by equally sharp decreases in production. As a result, if farm output at some time does not fall as much as market price,

21. Leonard W. Weiss, Economics and American Industry (New York, 1961), 62. Weiss cites a 1957 USDA study showing the main variable costs to be about 11 per cent of all costs on northern Plains spring wheat farms and 20 per cent on southern Plains winter wheat farms.

one cannot conclude that farmers behaved irrationally or that they would have gained from cutting output more than they did.

Returning to the situation faced by wheat farmers in the United States in the early twenties, one could argue that even though wheat acreage declined in the manner that our expectations about rational economic behavior would predict, wheat prices by 1924 might nevertheless have fallen too far for average producers to earn a fair return. This argument is in fact the main point of the conventional story about agricultural depression after 1920. Presumably there was excess capacity in American agriculture because of wartime expansion and the higher efficiency of new equipment available to farmers after the war. This excess capacity, together with reduced peacetime demands, meant that equilibrium farm commodity prices were below the average unit costs of most farmers, especially those raising wheat. Although much of the acreage added to wheat production during the war had been reduced when prices fell after 1920, and even though most wheat farmers still covered their variable costs at 1924 prices, few supposedly were able to cover all of their fixed costs which included a normal profit. This is the essence of the "overproduction thesis" which asserts that agricultural depression in the twenties resulted from too much output and unsatisfactory prices,

especially among commodity groups that undertook significant
expansion during the war.

If this "overproduction thesis" were correct,
however, and economic distress among wheat farmers after 1920
had been caused by low prices and excess capacity, one would
expect that the signs of economic distress would have been widely
distributed over the nation. Presumably changes in the market
price for any type of wheat should affect almost equally those
farmers who market the same variety and grade of wheat. Furthermore, unless wheat farmers in some areas expanded their
acreage during the war more than those in other areas, the
after-effects of wartime expansion should also have been wides-
presd. If wartime expansion were not widely distributed,
we would expect to find evidence of more severe postwar hardship in those areas where wartime changes in acreage had been
greatest. Thus, to test the "overproduction thesis" we must
determine what regional differences there were, if any, in
the degree to which wheat acreage expanded and contracted during
and after the war. The pattern of acreage adjustments must then
be correlated with evidence on regional differences in postwar economic distress.

Changes in wheat acreage planted in the United States during

World War I and in the early twenties are summarized in
Table IV - 2. Shown separately are changes in a group of
states where wheat-growing is largely a "specialized" commercial activity and few alternatives from the crop production
standpoint exist. The expansion of wheat acreage in those
states during the war would have to be considered more or
less irreversible.[22] In general, wheat grown in these states
is found in semi-arid regions, particularly in the Great
Plains. In the "other" states shown in the table, wheat
is usually grown where agriculture is more diversified and crop
alternatives to wheat are available. Clearly, the majority
of the acreage added during the war and all of the acreage
withdrawn from production just after the war was in the "other"
states, not in the specialized wheat states.

The argument that wartime expansion left many wheat
farmers stranded with excess capacity in the early twenties
is weakened by this evidence, since wheat was only one of
many crops grown in the states where acreage adjustments

---

22. A.E. Taylor, "The Dispensability of a Wheat Surplus in
the United States," 135.

TABLE IV - 2

NET CHANGE IN WHEAT ACREAGE PLANTED IN

THE UNITED STATES, 1912-1929

Net change in average acres planted ---	Specialized Wheat States ✱	All Other States	Total
	(thousands of acres)		
1912/1914 - 1918/1920	7,208	8,505	15,713
1918/1920 - 1923/1924	1,206	-7,741	-6,535
1923/1924 - 1928/1929	7,657	-595	7,062

✱Colorado, Idaho, Montana, North Dakota, Oklahoma, Oregon, Texas, Washington, and Wyoming.

Sources:
  1912 to 1924; A.E. Taylor, "The Dispensability of a Wheat Surplus in the United States," (1925), 142.

  1928 and 1929; USDA, Statistical Bulletin No. 158 (1955).

between 1914 and 1923 were greatest. The usual assumption
is that farmers opened new land when they expanded wheat
acreage during the war, and that once opened these acres could
only be sown with wheat or be abandoned. In reality, it appears
that much of the "new" acreage planted to wheat after 1914
was on land which previously had produced other crops, typically
corn. Referring again to Table IV - 2, almost 45 per cent of
the acres added to wheat in "other" states between 1912/1914
and 1918/1920 and over 35 per cent of the acres withdrawn in
those same states between 1918/1920 and 1923/1924 were in seven
Corn Belt states.[23] In these seven states, wheat is commonly
grown in rotation with corn. Thus, it is not surprising
that farmers in Corn Belt states shifted their acreage from corn
to wheat when wheat prices became more favorable than corn prices
during the war. It is also not surprising that farmers in these
same states were able to shift acreage back out of wheat when
wheat prices fell as sharply as they did after 1920.[24] The
horse-drawn technology still prevalent in midwestern American
agriculture in 1920 meant that such transitions were not as hard

23. Ibid., 142. The seven Corn Belt States are Ohio, Indiana,
Illinois, Michigan, Wisconsin, Iowa and Missouri.

24. This partially confirms the assertion made before that marginal cost functions, say for wheat, are less inelastic where
crop production alternatives of this sort exist. When wheat
prices fell, as they did in the early twenties, relatively
farther than corn prices, diversified producers could reduce

to make as in later periods when planting and harvesting machinery became much more specialized and costly.[25]

Problems faced by wheat farmers in the "specialized" (i.e. undiversified) states in Table IV - 2 were quite distinct from the problems that farmers in the "other" states faced. Although more acres were added to wheat in the "other" states during the war, a significant increase still did occur in the specialized states. Furthermore, as one would expect, there is little if any evidence of acreage withdrawn after the war in the specialized states. There, marginal cost would be far more inelastic so that a big drop in the price of wheat would not necessarily be followed by a large drop in production. Nevertheless, price may have fallen so low that specialized wheat farmers, unable to diversify, could not cover all fixed costs including normal profit, interest, and taxes. If this had been true, overexpansion, overproduction and low prices in the early

---

wheat output to a far greater extent than could specialized producers who had no viable alternative to wheat.

25. These comments apply of course to many areas outside the Corn Belt where wheat also was not a specialized crop, but where alternatives such as corn existed. In an interesting study published in 1925, W.E. Grimes of Kansas State Agricultural College noted that wheat acreage grew enormously during the war in the eastern one-third of Kansas where corn cultivation was prevalent. In central Kansas, however, which then was the heart of the Kansas hard winter wheat belt, wartime expansion was negligible. The reason, he noted, was that costs were too high during the war to warrant opening new acres in central Kansas while it was very reasonable to shift existing acreage in eastern

twenties would have caused many of the farm bankruptcies,
mortgage foreclosures and other signs of distress in those
specialized areas that were cited in "The Wheat Situation."

In fact, economic distress among wheat growers in the
early twenties appears to have been most severe in large
parts of the specialized wheat area discussed above. For
instance, we find near the end of "The Wheat Situation"
the following observation:[26]

> "Although financial difficulty is widespread
> among farmers in many regions where wheat is
> extensively grown, the situation is no doubt
> at its worst in the semiarid sections extending
> from western Kansas and eastern Colorado to the
> Canadian border. In these dry-land areas during
> the last few years farm indebtedness has grown in
> volume; delinquent farm interest and taxes have
> multiplied; and the capital and credit of farmers
> have been so depleted that it has been necessary
> to provide county, State, and Federal funds for
> seed and feed loans."

In spite of these remarks, however, it was true that economic
distress was not evenly distributed among the parts of this
Great Plains wheat-growing area. Failure to distinguish
between economic conditions in northern and southern sections
of the Plains was one serious weakness of the analysis in "The
Wheat Situation."

Kansas from corn to wheat. See W.E. Grimes, "Some Phases of
the Hard Winter Wheat Grower's Problem in Readjustment," Journal
of Farm Economics, Vol. VII, No. 2 (April, 1925).

26. USDA, "The Wheat Situation," 143.

The section of the Great Plains that is referred to in
the above quotation is without question the most concentrated
and single most important wheat-growing region of the United
States. Within that region, however, there are great differences
in climate and type of wheat produced. Basically, northern
areas of the Plains specialize in hard spring wheat while
southern areas specialize in hard winter wheat.[27] To simplify
our analysis, North Dakota can be studied as a typical northern
Plains spring wheat area while Kansas can be considered a
typical southern Plains winter wheat area. Spring and winter
wheats of several varieties are of course grown in many other
states both in the Plains and elsewhere in the United States.
Nevertheless, the conclusions that were reached in "The Wheat
Situation" can be adequately tested by comparing conditions
that prevailed in North Dakota and Kansas.

Several types of evidence exist to show that economic
conditions were far worse in the early twenties in northern
spring wheat areas like North Dakota than in southern winter
wheat areas like Kansas. One piece of evidence noted at the

27. An excellent discussion of the different varieties of
wheat grown in the United States in this period (and later),
including details on regional distribution of varieties and
acreage, is in Jeanne E. Dost, An Interregional Analysis
of the Three Major Wheat Producing Regions of the United
States, unpublished Ph.D. thesis (Radcliffe College, 1958),
chapter I.

time was the high rate of farm bankruptcies that began to
plague North Dakota after the war. Farm bankruptcies in North
Dakota rarely numbered more than 60 to 70 per year before
1922 when they shot up to 237 and then to more than 600 and
700 in each of the three following years. In Kansas, on the
other hand, where pre-1920 experience had been only slightly
better than in North Dakota, bankruptcies also rose markedly
from 1921 to 1925, but the number was not much above 225 per
year.[28] The lower absolute number of bankruptcies in Kansas
was striking since more than twice as many farm operators were
listed by the 1920 Census in Kansas than in North Dakota.[29]
Similarly, farm mortgage foreclosure statistics reflected worse
conditions in North Dakota than in Kansas, although these data
are available only after 1925; however, foreclosure figures
would reflect cumulative problems from prior years as well
as current financial difficulties. Thus, from 1926 through
1929, in central and eastern sections of North Dakota where
specialized wheat farming was dominant, an annual average of

---

28. David L. Wickens, Farmer Bankruptcies, 1898-1935, USDA,
Circular No. 414 (Washington, D.C., September, 1936), 5-6.

29. U.S. Department of Commerce, U.S. Census of Agriculture:1950,
Vol. II, Table 22, 996.

from 2.9 to 5.4 per cent of all farms were transferred through
mortgage foreclosure. In the same period, average annual foreclosure rates in the wheat specialty sections of central and
western Kansas ranged from .7 to 3.4 per cent.[30] Finally, the
average annual percentage of bank deposits in suspended banks from
1920 to 1929 ranged from 3.4 to 7.1 per cent in the North Dakota
wheat sections and from .7 to 1.8 per cent in the Kansas wheat
areas.[31]

There is also evidence less direct but perhaps more significant than statistical data which leaves no doubt that contemporaries were aware that northern spring wheat farmers suffered greater
hardship than their counterparts in southern winter wheat areas
after World War I. Recognition of this was implicit in the
tours through distressed wheat states made in late 1923 by
Eugene Meyer, who represented President Coolidge, and Henry
C. Taylor, who represented Secretary of Agriculture Henry C.
Wallace. Although Meyer and Taylor represented opposing camps in
the Washington battle over farm legislation, both men restricted
their tours to northern spring wheat states and the Pacific

30. Foreclosure rates by "crop reporting area" were compiled
from worksheets supplied by USDA. These worksheets were the
ones used to prepare data for The Farm Real Estate Situation,
a bi-annual bulletin USDA began publishing in the late twenties.

31. Bank suspension data compiled from worksheets supplied
by the Federal Deposit Insurance Corporation. These worksheets,
prepared in a WPA project during the late thirties, are on file
at the Graduate Program in Economic History, Madison, Wisconsin.

Northwest; neither man visited any southern winter wheat states.[32]
Also striking were comments by two North Dakota agricultural
economists, John Coulter and Rex Willard, who agreed that
postwar economic difficulties faced by wheat farmers in their
state were far more severe than any problems faced by farmers in
winter wheat areas.[33]

Coulter's and Willard's comments bring us in fact
to the heart of the main flaw inherent in "The Wheat Situation."
As suggested in "The Wheat Situation" and accepted by historians
ever since, overproduction and low prices caused the problems
faced by wheat farmers on the Plains during the early twenties.
If so, then why were these problems not more widespread? Obviously,

---

32. J.H. Shideler, Farm Crisis: 1919-1923, 258-259 and 262-263.
As Shideler noted, Taylor's itinerary included visits to southern
Plains states; however, President Coolidge supposedly asked
Secretary Wallace to recall Taylor to Washington before his
trip was complete. Wallace relayed orders to return to Taylor
on October 24, 1923. In spite of this, Taylor had written
Wallace a week earlier (October 16th) asking permission to return
ahead of schedule, apparently convinced that he had seen enough
farm "distress" to understand the problems at hand. A separate
file of correspondence and other material on Taylor's tour, including his observations on the Meyer mission, is in the Henry C.
Taylor papers, Box No. 2, State Historical Society of Wisconsin,
Madison.

33. John Lee Coulter, "The Wheat Crisis," The Quarterly
Journal of the University of North Dakota, Vol. 14, No. 1.
(November, 1923), 3-26. Rex E. Willard, "Comments," Journal of
Farm Economics, Vol. VII, No. 2 (April, 1925), 220-221.

changes in wheat prices affect all areas about equally --
especially when the price change being considered was a drop
as severe as the one in 1920/1921. Nevertheless, North Dakota
wheat farmers faced a far more serious setback than did Kansas
wheat farmers in the early twenties. It would seem that unfavorable factors other than overproduction and low prices were
at work in North Dakota at that time.

Indeed, other factors did influence the fortunes of North
Dakota wheat farmers. Basically, the effect of these other
factors can be summed-up in two words: low yields. For one thing,
northern spring wheat areas in the United States suffered severe
drought from about 1917 to 1921, a point noted in "The Wheat
Situation." Other wheat-growing areas of the Plains escaped
drought at that time.[34] The severity of this drought prompted
federal legislation to give stricken farmers seed-grain loans;
however, almost all of the loans were distributed in North Dakota

34. USDA, "The Wheat Situation," 120. This drought also affected
yields in wheat-growing sections of Manitaba and Saskatchewan just
north of Montana and North Dakota. If Henry Taylor had not realized
how great the impact of low yields was in North Dakota during his
October, 1923 tour, the point was made clear to him in an eight-
page letter from John Coulter dated February 26, 1924. Using
USDA data on annual wheat yields in N. Dakota since 1882, Coulter
noted that decreased yields was "one of the greatest causes for
our distress" and that such a fact was "not a political question
but purely an economic and agricultural one." (letter, page 3)
Filed in Henry C. Taylor Papers (Box No. 6, "John Lee Coulter,
1907-1932"), State Historical Society of Wisconsin, Madison.

and Montana.[35] In addition to drought in the early twenties,

spring wheat areas like North Dakota suffered throughout the

decade from blight, mainly in the form of stem rust.[36] Therefore,

the combined effect of drought and blight served to reduce

acreage yields in North Dakota to levels well below those in

Kansas during most of the 1920's. Low yields, especially when pro-

longed over several crop years, could obviously cause severe

economic distress in an area like North Dakota where crop

production alternatives were limited. Wartime expansion,

overproduction, low prices, and all the other factors to which

the hard times of the early twenties are generally traced

have little relevance when rain does not fall and where bugs des-

troy much of the crop that does manage to sprout.

It seems unlikely that northern spring wheat farmers, at

least in North Dakota, would have benefited much if the price

of wheat could have been raised in the early twenties. In

---

35. Ibid.

36. The high incidence of rust in spring wheat areas was
noted by Rex Willard in his article cited above. Also it was
noted by Jeanne Dost in her thesis, An Interregional Analysis
of the Three Major Wheat Producing Regions of the United States,
chapter II.

retrospect, the price-support scheme endorsed in "The Wheat
Situation" was not the solution to their grievance. Nevertheless,
historians have accepted the analysis and conclusions put forth
by Henry C. Wallace, Henry C. Taylor and other members of the
USDA in that bulletin. Few scholars have viewed the agricultural
depression of the early twenties in the light of evidence put
forth by other analysts of the time, such as Coulter, Willard
and Grimes. Coulter and Willard both opposed legislative schemes
to raise wheat prices simply because they did not feel price
was the culprit in North Dakota. The arguments presented above
suggest, however, that the traditional emphasis historians place
on prices and overproduction may be mistaken. Wheat prices of
course fell drastically in 1920/1921, but our analysis revealed
that the nation's wheat farmers adjusted their acreage in the
wake of this price collapse. The adjustment was almost completed
by 1924. Furthermore, a comparative analysis revealed that
economic distress in the early twenties was much less evident
in the areas that experienced most of the acreage adjustment.
Severe distress was found where wheat farmers faced problems

caused by drought and blight. Thus, in light of the above
discussion, the central role which economists and historians
have given to overproduction as the cause of farm distress
in the early twenties appears questionable.

Nevertheless, historical and popular accounts, following
the argument of "The Wheat Situation," have retained overproduction
and low prices as the main causal factors for agricultural depres-
ssion after 1920. A possible reason for this lasting emphasis is
that wheat prices fell again in the late twenties and, so it
seems, because of continued overproduction. Proof for this
assertion is often drawn from evidence that wheat stocks on
hand in the United States in mid-1929 were greater than ever
before for that time of year. Furthermore, wheat prices had been
falling again since 1925, and remained low in 1929 while these
stocks were accumulating.[37] Such data have suggested to many
scholars that warnings raised in "The Wheat Situation" in late
1923 about the long-run possibilities for overproduction

---

37. Data on wheat stocks before 1928 are in Food Research
Institute, "Disposition of American Wheat Since 1896," Wheat
Studies, Vol. IV, No. 4 (February, 1928), 180. Stocks after
1928 are in U.S. Department of Commerce, Historical Statistics of
the United States, 296:K-273. Wheat price data are in F.L.
Strauss and L.H. Bean, Gross Farm Income and Indices of Farm
Production and Prices in the United States, 1869-1937, 36, Table 13.

were perhaps correct.

Nevertheless, data on national wheat stocks are an
ambiguous index of surplus production. Low levels of wheat
stocks usually reflect that no surpluses exist; however, high
stock levels in a period like the late twenties could have
reflected easy credit and hedging by wheat processors against
inflation rather than overproduction of the commodity. Coupled
with this was the coincidence in 1929 of a very favorable wheat
crop in both the United States and Argentina. The latter nation,
with much poorer storage facilities than the United States, flooded
British markets with an unprececented amount of wheat in 1929 and
thereby dampened world wheat prices.[38] Another factor affecting
American wheat stocks in 1929 was the enactment of the long-
awaited Federal Farm Board under the Agricultural Marketing
Act of 1929. Many farmers and sepculators held unusually high
stocks of wheat, anticipating that the wheat board would act
to raise prices.[39] Thus, the high level of wheat stocks in

---

38. Food Research Institute, "The World Wheat Situation,
1928-29," Wheat Studies, Vol. 6, No. 2 (December, 1929), 66ff.

39. Ibid., 58 and 67.

1929, instead of being a sign of overproduction, reflected some short-run phenomena, the effects of which might well have vanished in a year of two had the world market not fallen apart in 1930.

Nevertheless, wheat acreage was again increasing in the United States after 1926 and wheat prices were falling. Was this increase in acreage a resurgence of underlying malad-justments set in motion by wartime expansion and accelerated by the increased efficiency of equipment coming on the market after 1920? There is no question that farm distress remained a major issue in the late twenties. Was this because farmers were again faced with overproduction and low prices? A definitive answer cannot be offerred at this time; however, a brief look at the areas where most of the expansion of wheat acreage in the late twenties occurred will shed some light on the question.

It is clear from Table IV - 2 that all of the expansion in wheat acreage after 1923/1924 occurred in specialized wheat-growing areas. The state data from which this table was compiled reveal that over one-half of this expansion in spec-

ialized areas was in winter wheat districts of the Pacific
Northwest and semi-arid sections of the Southern Plains.
In fact, almost two-thirds of the net increase in acres planted
in specialized wheat states was in winter wheat areas, if
Kansas is included in the specialized states instead of the "other"
states. Wheras most of the wheat acreage adjustments in Kansas
between 1912/1914 and 1923/1924 were in central and eastern
counties where conditions are similar to those in the "other"
states, the expansion of acreage after 1923/1924 occurred
mainly in the semi-arid western one-third of the state where
wheat-growing became a highly specialized enterprise.[40] Once
again, therefore, a comparison of conditions in Kansas and
North Dakota can shed light on the forces behind and probable
effects of wheat acreage expansion in the late twenties.

On the demand side, several conditions in the late
twenties favored acreage expansion in Kansas winter wheat
rather than North Dakota spring wheat. One factor was the
rise of Kansas City as a major milling center after the early
twenties. This narrowed some of the competitive advantage

---

40. W.E. Grimes, "Trends in the Agriculture of the Hard Winter
Wheat Belt," The Journal of Land and Public Utility Economics,
Vol. IV, No. 4 (November, 1928), 347-354.

which North Dakota farmers had long enjoyed by virtue of their proximity to Minneapolis.[41] Another factor favoring Kansas growers was the continued presence during the twenties of a long-standing price-bias against hard winter wheats in comparison with hard spring wheats. Traditionally, spring wheats were more favorably priced than winter wheats, although it was apparent by the 1920's that the baking qualities of either variety were the same. During the twenties, however, the continuing transfer of baking activities, especially of bread, from the home to commercial bakeries became more pronounced than ever before. Commercial bakers, aware of the high quality of hard winter varieties and even more aware of their relatively lower price, generated an enormous demand for the output of southern Plains wheat-growing areas like western Kansas.[42]

On the supply side, conditions were even more favorable for wheat acreage expansion in western Kansas. One reason for this was of course the low yields from rust in northern

---

41. J.E. Dost, <u>An Interregional Analysis of the Three Major Wheat Producing Regions of the United States</u>, chapter II.

42. <u>Ibid</u>., chapter I. William G. Panschar, <u>Baking in America</u> (2 vols., Evanston, Illinois, 1956), Vol. 1, 99.

spring wheat areas like North Dakota. Another factor,

perhaps more important, was the advent of the tractor and

tractor-driven combine after 1920. In 1920 an estimated 4,000

combines were in use on American farms, but by 1928 this number

had grown to over 45,000. Most of those combines in 1928 were

on southern Plains farms, for example 20,000 in Kansas, 6,000

in Oklahoma, and 3,000 in Texas. Only 2,000 were in North

Dakota.[43] A study done in the late thirties showed that

only 5 per cent of the spring wheat farms around North Dakota

reported using combines in 1929, while 86 per cent of the

winter wheat farms in western Kansas reported using combines at

the same date.[44]

Different climatic conditions at harvest time in North

Dakota had much to do with the lower rate of adoption of combines

there in the 1920's. In North Dakota, at that time, harvesting

operations could not begin until September or later. In

western Kansas, however, winter wheat harvesting could be conducted

---

43. M.R. Cooper, et. al., Progress of Farm Mechanization, 32.
U.S. Works Projects Administration, Changes in Technology and
Labor Requirements in Crop Production: Wheat and Oats (Report
No. A-10, Philadelphia, April, 1939), 133.

44. Ibid., 37 and 43.

in July and August. As a result, the wheat at harvest time
tended to be much drier in Kansas than in northern spring wheat
areas, which made combining more feasible. If wheat is moist
when it is combined, chances of loss from smut are significantly
increased. Thus, wheat growers in North Dakota often had no
choice but to rely on headers and binders. They could not
wait in most cases until the wheat was dry enough to combine,
since they then took the risk of losing the entire crop to wind
or rain. Kansas producers, on the other hand, rarely faced
such risks during the hot, dry summer harvest season.[45]

The significance of the combine to western Kansas wheat
farmers in the twenties is shown by evidence that man-hours
per acre to raise wheat in that general area fell from 5.6
hours in 1919 to 2.6 hours in 1929. This labor-saving was
comprised of a reduction in preharvest labor time from 2.2 to
1.3 hours and in harvest time from 3.4 to 1.3 hours. On the
other hand, man-hours per acre in northern spring wheat areas
around North Dakota fell from 5.6 hours in 1919 to only 5.2
hours in 1929. There, preharvest time was reduced from 2.6
to 2.3 hours and harvest time from 3.0 to 2.9 hours.[46] As a

45. American Society of Agricultural Engineers, Present Status
of "Combine" Harvesting (St. Joseph, Michigan, March, 1928).

46. U.S. WPA, Changes in Technology (A-10), 37 and 43.

result, one economist has estimated that average total cost
per bushel in 1929 was $.618 in hard red winter wheat areas
and $.734 in hard red spring wheat areas. Mainly because
of continued mechanization, but also because of improved
yields, the same figures by 1939 were $.357 and $.612,
respectively.[47]

Therefore, shifts in the industry supply function
probably explain much of the decline in wheat prices after
the mid-twenties. Since these shifts in supply were mainly
due to labor-saving mechanization rather than land-saving
yield increases, they resulted in acreage expansion. The rise
in acreage and output that accompanied falling wheat prices
in the late twenties was not, therefore, a sign of industry
overproduction. Industry overproduction implies that market
price is too low for the average firm in the industry to cover
all costs of production. In many wheat areas of the United
States in the late twenties, however, we saw that costs were
falling as well as prices. This was especially true of western
Kansas. In fact, farm land value statistics suggest that net
profits increased in that area throughout the late twenties.
In the western one-third of Kansas, farm land values rose

47. J.B. Sjo, Technology: Its Effect on the Wheat Industry,
Table 43, 190.

steadily from 1927 to 1932.[48] On the other hand, conditions

in North Dakota, although better than in the early twenties,

were not as favorable in the late twenties as in Kansas. Many

more North Dakota wheat farmers suffered bankruptcy and foreclo-

sure as prices fell. Nevertheless, if these signs of distress

reflected falling net incomes, the main culprit does not

seem to have been falling prices. Rather, the problem arose

because the North Dakota producer was unable to cut costs

as rapidly as price fell. The main choice for North Dakota

wheat farmers was probably to produce something besides

wheat or else leave farming.[49]

In summary, the discussion here suggests that concepts

like "overproduction" and "parity" and a simplistic emphasis

on price data are inadequate to depict the range and depth

of economic problems faced by American farmers in the 1920's.

In wheat-growing areas at least, problems in the early twenties

---

48. Taken from farm land value data by "crop reporting" district,
supplied by USDA. Also see E.H. Wiecking, The Farm Real Estate
Situation, 1929-30, USDA, Circular No. 150 (Washington, D.C.,
November, 1930), 37.

49. The authors of "The Wheat Situation" saw price supports
(through export dumping) as an alternative choice; however, it
is not clear that this would have solved anyone's problems. If
wheat prices had been supported above market levels in the late
twenties, low-cost producers in Kansas would have expanded
acreage even more than they did. This would have put downward

that were attributed to surpluses and low prices in "The Wheat

Situation" appear on closer analysis to have been due to natural

adversities such as drought and blight. The effects of wheat

acreage expansion during World War I have been misunderstood by

most historians and economists who write about agricultural

depression in the 1920's. Furthermore, the effects of acreage

expansion and falling wheat prices in the late twenties have

also been misunderstood or, in some cases at least, ignored.

The wheat industry was not then facing overproduction as much

as it was experiencing the impact of technological change and

shifting patterns of resource use. Technological change and

resource adjustment have not, however, been popular themes in

most accounts of agricultural distress in the 1920's. Perhaps

the reason these themes are so often ignored is that so little

data is available to study them. It has been easy for historians

---

pressure on prices, creating the need for massive dumping
operations. The "solution" to this problem adopted after 1933,
namely, voluntary acreage limitation and subsidies, has tended
to distort efficient allocation of resources in wheat farming
by encouraging production in high-cost areas. This last point has
been amply demonstrated in J.E. Dost, An Interregional Analysis
of The Three Major Wheat Areas of the United States, chapter III.

and economists to fall back on the aggregate averages that are
supplied so abundantly by the U.S. Department of Agriculture.[50]
Likewise, it is understandable that users of such data generally
adopt the analytical schemes from which these data originate
and tend to support. In this context, it is important to remember
that statistical data are not autonomously generated and do not
necessarily reflect reality _per se_. Quantitative data exist
because some one derived them, not because someone just found
them. Parity statistics were derived because of an idea,
not because they show a fact of life. Conversely, data like
parity price series often have a tendency to reveal exactly
what the ideas behind them suggest they should reveal. In any
event, we have a plethora of data on farm commodity prices,
parity ratios, production averages, and national income. A
major research effort will be necessary, however, before we
can assess accurately why some farmers failed in the twenties
and why some prospered. Our discussion of wheat farmers shows

---

50. One group of agricultural engineers, interested in ways
to increase the productive efficiency of American farms in the
twenties, found it difficult to derive any useful information
on farm productivity from the type of data supplied by USDA.
They urged that "...in the compiling of agricultural statistics
and data, major importance be given to results obtained by the
best agricultural practices, rather than to average results."
A Conference of the American Society of Agricultural Engineers
with Calvin Coolidge in Rapid City, South Dakota, mimeograph
(August 30, 1927), 20. Filed in the National Agricultural
Library, USDA, Washington, D.C.

how little is known about specific cases.

These four chapters have revealed flaws in the traditional story of agricultural depression in the 1920's. In particular, the analytical scheme on which this traditional story rests has been challenged. Nevertheless, falling farm land values as well as high farm mortgage foreclosure and rural bank failure rates after World War I have convinced many scholars that farmers did face severe economic difficulties in the twenties. Up to this point, our analysis has suggested that the alleged causes for this distress have probably been erroneous. Except for the discussion of wheat conditions, the actual picture of economic distress itself has not been considered. Thus, we attempt in the next chapter to find reasons for such things as the high failure and foreclosure rates among American farmers after World War I.

CHAPTER V

WARTIME OPTIMISM AND FARM CRISIS IN THE TWENTIES:
LAND VALUES, MORTGAGES AND BANKS

Material in the preceding chapters casts doubt on much
of the conventional wisdom regarding agricultural depression
in the 1920's. Income and parity data, when subjected to
close analysis do not confirm the notion that farmers as a group
failed to share in the general prosperity of the twenties.
Furthermore, a study of conditions among wheat farmers following
World War I does not support the thesis that "overproduction"
was to blame for economic distress in agriculture. Economic
hardship of course existed in certain areas where wheat pre-
dominates; however, the causes of this hardship were never
the usual ones cited as reasons for agricultural depression in
the twenties.

Historical literature abounds with other evidence,
besides that relating to wheat areas, showing that all was
not well for many American farm operators after 1920. For

one thing, average farm land values declined steadily in almost
all states after 1920, reversing the rising trend that had been
taken for granted in most rural areas before the war. Farm
mortgage foreclosure rates also rose to high levels and became
a critical and persistent problem in many farm areas during
the postwar decade. Finally, although most parts of the
United States, rural and urban, experienced a high number of
bank failures after 1920, the number of failures in rural areas
was very alarming.

The trends in each of these factors -- land values,
mortgage foreclosures, and bank failures -- have usually been
taken to reflect farm economic depression in the twenties.[1]
Furthermore, the similarity of trends in these factors in the
twenties and early thirties seems to support the notion that
economic conditions on the farm were similar in kind if not
in degree during both periods. Therefore, each of the above
three indices of economic distress in the 1920's will be examined
in this chapter to determine what forces lay behind them. The

---

1. An excellent example of how historians use these data to
confirm the existence of agricultural depression in the 1920's
is Theodore Saloutos and John D. Hicks, Agricultural Discontent
in the Middles West, 1900-1939 (Madison, Wisconsin, 1951), 100-105.

basic problem to be considered is whether the trends shown
during the twenties in such data can either explain or be
explained by the supposed existence of agricultural depression
after World War I.

A. Farm Land Value

The persistent and pervasive decline in the market
value of farm real estate after 1920 is often viewed as a
sign of economic distress in agriculture during the twenties.
Rising land value, urban and rural, had been a feature of
American economic life from the time of the earliest settlements;[2]
however, opposite trends in the value of farm and city real
estate in the 1920's reinforce the impression that prosperity
in that decade was not enjoyed by the rural sector of the
economy. In fact, there was a continuous decline in the average
value of farm land in all but a few states from 1920 to the
late thirties, which appears to support the view that American
agriculture was beset with chronic economic difficulties for
almost two decades after World War I.[3]

2. Aaron M. Sakolski, The Great American Land Bubble (New York,
1932), passim.

3. U.S. Department of Agriculture, Yearbook of Agriculture; 1931
1025, Table 540.

A long-term decline in farm land values certainly
would have been cause for concern in any period, since capital
gains had always been an important component of lifetime income
for the average American farmers.[4] Throughout the ninteeenth
century, the rate of turnover was high in most farming districts
during the first generation of settlement. Only as tenure
became stabilized did one find farmers who paid closer attention to costs of operation than to land profits.[5] Nevertheless,
the steady rise in commodity prices and farm land values from
1900 to 1915 strengthened the impression of many farmers,
especially in the Midwest and West, that farm land bought at
any price in one period could always be sold for a higher
price in a later period. Therefore, an expectation which
farmers had taken for granted for many years was dampened by
the steady fall in values after 1920.[6]

---

4. Allan G. Bogue, From Prairie to Corn Belt (Chicago, 1963),
266 and 287. James C. Malin, The Grassland of North America:
Prolegomena to its History (Lawrence, Kansas, 1947), 295ff.

5. Paul W. Gates, The Farmer's Age: Agriculture: 1815-1860
(New York, 1960), 399-403. James C. Malin, "Mobility and History,"
Agricultural History, Vol. 17, No. 4 (October, 1943).

6. Expectations of long-run growth in farm land values were
not completely dampened after 1920. One group (M.L. Wilson
of Montana State Agricultural College was included) attempted
to form a sort of "mutual fund" in the early twenties that
would invest subscribers' funds in midwestern farm properties.
As the prospectus of the Northwest Land and Finance Corp. of
St. Paul, Minnesota pointed out, "the opportunities for profit
through investment in the most desirable of selected improved
farms, are exceptional. They have not existed in the same degree

It is noteworthy that an enormous boom in farm land
values occurred in many parts of the United States during 1919
and early 1920. Probable causes and effects of this sharp
rise in values will be discussed throughout the following
pages; however, the steep descent in average farm land values
after 1920 cannot be properly appreciated without considering
the postwar boom. Since many scholars have written illumin-
ating accounts of the land market mania that gripped farm
areas just after the war, only scant attention will be given
to that matter here.[7] In one account of agricultural depression
in the twenties and early thirties, however, Theodore Saloutos
and John D. Hicks said that "probably the factor which contributed more than any other to the deepness of this depression
was the land boom that had accompanied the war prices."[8] Thus,

---

during the past twenty years, and probably will not exist
again. Lands thus acquired can be carried at a profit until
the appropriate time comes to liquidate the investment."
A copy of the prospectus is in U.S. National Archives, Files
of the Bureau of Agricultural Economics, Box 105, "Farm
mortgage foreclosures."

7. G.E. Mowry, The Decline of Agriculture, 1920-1924; unpublished
M.A, thesis (University of Wisconsin, 1934), chapter 1. L.J.
Norton, "The Land Market and Farm Mortgage Debts, 1917-
1921," Journal of Farm Economics, Vol. 24, No. 1 (February, 1942),
168-173. T. Saloutos and J.D. Hicks, Agricultural Discontent
in the Middle West, 101-105.

8. Ibid., 101.

the drop in values during the twenties was steep in relation
to the high levels reached by early 1920, but in no region
of the United States were average farm land values lower in
1929 than they had been before the war (Table V - 1).

The fall in land values during the twenties probably
affected farm operators differently, depending on what they
did during the postwar land boom. Those who purchased farm land
between 1919 and 1920 or who incurred mortgage debt that was
secured by land valued at prices during that period surely
saw their equity wither away during the twenties. No matter
how profitable their farm operations may have been after 1920,
these farmers no doubt faced bankruptcy or foreclosure solely
because of shrinking land values. Other farmers who acquired
neither land nor debt during the postwar boom would not have
been affected so much by the slump in land values after 1920.
Unfortunately we have insufficient evidence to determine
exactly how many farmers purchased or secured land during the
postwar boom. One limited study of land transfers in Iowa during
1919 suggests that only 10 per cent of all farms changed
hands during the most intense period of postwar land activity.[9]
More will be said later about mortgage transactions during that

9. L.C. Gray and O.G. Lloyd, Farmland Values in Iowa, USDA,
Bulletin No. 874 (Washington, D.C., 1920).

TABLE V - 1

FARM REAL ESTATE: INDEX NUMBERS OF VALUE PER ACRE, UNITED STATES AND CENSUS REGIONS, 1912/1914 - 1929.

	US	N Eng	M Atl	ENC	WNC	S Atl	ESC	WSC	Mtn	Pac
1912/1914	100	100	100	100	100	100	100	100	100	100
1915	103	99	100	104	105	98	99	100	98	107
1916	108	102	104	110	114	108	109	103	98	111
1917	117	112	112	116	122	119	120	116	106	122
1918	129	117	117	127	134	135	140	134	117	129
1919	140	123	121	135	147	161	162	143	130	134
1920	170	140	136	161	184	198	199	177	151	156
1921	157	135	127	151	174	174	163	159	133	155
1922	139	134	118	132	150	146	149	136	122	151
1923	135	130	116	128	142	152	149	132	115	148
1924	130	128	114	121	132	151	142	136	110	147
1925	127	127	114	116	126	148	141	144	105	146
1926	124	128	113	111	121	149	139	144	103	144
1927	119	127	111	104	115	137	133	139	101	143
1928	117	127	110	101	113	134	130	137	101	142
1929	116	126	109	100	112	132	129	136	101	142

Source: USDA, Yearbook of Agriculture:1931, 1025, Table 540. Value estimated at March 1 each year. The census regions are: New England, Middle Atlantic, East North Central, West North Central, South Atlantic, East South Central, West South Central, Mountain, and Pacific.

era; however, it is not obvious that the post-1920 drop in
farm land values was a major cause for economic distress during
that decade, nor does it seem likely that the fall in values
reflected current operating conditions during the 1920's.

Most writers, however, do not seem to attribute the
problems caused by falling farm land values in the twenties to
the fall in land value itself. Rather, they cite the fall in
values as an index of underlying economic weaknesses in the
agricultural sector. In other words, it is alleged that farm
land values continued falling after 1920 mainly because farm
prices and incomes had fallen. Behind this reasoning is the
principle that the long-run market value of land, as of any
asset, is the discounted value of an expected stream of annual
earnings from the land. Therefore, if land income is falling
then land values will fall, provided the rate of discounting remains
stable.[10] A problem arises, however, when this basic principle
of asset valuation is applied in reverse, as it is when hist-

10. Annual data on mortgage interest rates recorded between
1910 and 1929 reflects very little change in these rates from
one year to another in any region of the United States. See
D.C. Horton, H.C. Larsen, and N.J. Wall, Farm-Mortgage Credit
Facilities in the United States, USDA, Misc. Publication No. 478
(Washington, D.C., 1942), 229-231.

orians assume that falling farm land value automatically reflects
falling farm income. This assertion ignores the sharp rise in
values in 1919 and 1920 which perhaps influenced post-1920 farm
land values more than current operating income.

The income data discussed in chapter II revealed that
average farm incomes per capita in the United States rose
almost without interruption from 1921 to 1929. Income data
on a state or commodity-area basis are not available for that
period; however, regional estimates of farm net incomes prepared
by USDA for years after 1922 suggest that each census region
in the United States shared in the national growth of farm
income during the twenties. In some states like North Dakota,
Montana, and Georgia, adversities of nature severely affected
farm incomes early in the decade, but in most other states average
farm incomes no doubt followed the trend shown in the national
statistics. Therefore, it seems unlikely that depressed incomes
after 1921 could explain the fall in farm land values during
the twenties.

Nevertheless, there is one sense in which the trend in
farm incomes after 1920/1921 probably did influence farm land
values. Theoretically, farm land value should equal current

land income,for which land rents can serve as proxy, capitalized
at the current discount rate, usually the local mortgage interest
rate. This simple relationship between land income and land
value can break down, however, if land income (rent) has been rising
for a long period of years, as was true between 1900 and 1920.
In that case land value can become the capitalized value not
just of current income but of expected future increases in income
as well. The fall in average farm land values after 1920, then,
might have reflected that farm incomes during the twenties
failed to rise as fast as was necessary to justify the high
prices paid for farm land in 1919 and early 1920.

One author, in a study of 1920 farm land values in the
United States, estimated the importance of expected annual
increases in land income that would have been required in subsequent years to maintain the land values prevailing in 1920.[11]
Assuming that land incomes were capitalized at current mortgage
interest rates, Clyde Chambers analysed data on farm land
values and rents in 1920 that were available for certain sections
of the nation. Chambers found, for instance, that 50 per cent

11. Clyde R. Chambers, Relation of Land Income to Land Value,
USDA, Bulletin No. 1224 (Washington, D.C., June, 1924), The
following references are from an article drawn from Bulletin No.
1224 that was published in American Economic Review (December,
1924), 673-698.

or more of the value of farm land in much of the Middle West
was based on expected future increases in land income. Only
50 per cent or less of total land value was attributed in such
cases to the discounted value of **current** land income.[12] For
Iowa, the one state he examined for a period of years before 1920,
Chambers found that farm land values were capitalized from
expected annual rent increases which themselves had increased,
gradually from 1910 through 1918 and then very sharply between 1919
and 1920. By 1920, current farm land values in much of Iowa
were based largely on the expectation that land incomes (rents)
would not only continue rising, but would continue rising
at an accelerated rate.[13]

    Such expectations were no doubt fulfilled for a short
while after late 1918;[14] however, unless accelerated growth in

---

12. Ibid., 681, Table 3, Groups 7, 9, 10, and 14. Chambers
used the discounting formula $V = a/r + i/r^2$ where V is current
land value, a is current land income (rent), r is the rate of
capitalization, and i is the expected annual increase in income
for the next year which is necessary to justify V. Thus, Chambers
took i/r^2 divided by V as the portion of V made up of expected
future increases in land value.

13. Ibid., 686, Table 5, Column 5. Holding r constant
at 5.5 per cent, Chambers showed that average V in Iowa was
based on a value for i of $.06 in 1910, $.16 in 1916, $.18 in 1918,
$.19 in 1919 and $.32 in 1920.

14. A survey in Iowa in 1919/1920 persuaded officials of the
Office of Farm Management, USDA, that farm land values were
based on current farm commodity prices and the expectation
that recent
upward trends would continue for ten years at least. See O.G.
Lloyd, "Studies of Land Values in Iowa," Journal of Farm Economics,
Vol. 2, No. 3 (July, 1920). For the sequel to this story see

incomes could have been maintained after 1920, it was inevitable
that farm land values would drop sharply, as they did, after
that date. The expectation of accelerated increases in farm
income was based on unusual demand and price conditions during
the period of war and postwar European reconstruction. To have
counted on such conditions prevailing for a long period after
1919 can hardly be termed anything but "speculative."

The severity of the drop in farm land values after 1920
can therefore be attributed in large part to speculative overvaluation of land in 1920.[15] After the early twenties, the
rising level of current farm incomes no doubt became the dominant
factor once again in determining land value trends. This is
evident when one notes that farm land values, although continuing

---

Harry R. O'Brien, "Iowa's Abandoned Farms: A Study of How the
Inflated, Speculative Prices of Land Affect the State," The
Country Gentleman (June 18, 1921), 8 and 28.

15. Rough evidence to back this assertion is provided by correlating
the per cent rise in farm land value from 1918 to 1920 with the
per cent fall in value from 1920 to 1923 for the nine census
regions of the U.S. in Table V - 1. The land boom did not take
hold until late 1918 in most areas, while by 1923 the worst
of the crash was over. Simple rank correlation of the above
per cent changes yields an r coefficient of .717 which is significant at the 5 per cent level. A least-squares linear regression
yields an R^2 coefficient of .564, which is significant at the
5 per cent level. On this evidence one could say that over one-
half of the fall in farm land values from 1920 to 1923 is "explained" by the rise in values from 1918 to 1920.

to fall in all regions of the United States during the twenties, fell at a decreasing rate after 1922.[16] In fact, by 1929 average farm land values were moving along a stable horizontal trend in most regions. Rising farm incomes had in most instances dampened the extent of the readjustment from postwar overvaluation. If the world-wide depression of 1930 had not intervened, it seems probable from what was said before that farm incomes would have continued rising after 1929. In that case, farm land values would undoubtedly have risen after 1929, as they had for many years up to 1920.

Therefore, it is difficult to attribute the decline of farm real estate values after 1920 to the alleged agricultural depression of that decade. More than anything, the fall in land values seems to have been an inevitable adjustment to unreasonable expectations and overvaluation during the land boom of 1919 and early 1920. Many farmers surely suffered capital losses during the twenties, especially if they had bought land immediately after the war; however, this was not the case for most farm operators. In any event, the subsequent losses of those who did buy during the early postwar years had little to do with general trends in agricultural prices, production or

---

16. This had been noted by John D. Black in 1929 in his *Agricultural Reform in the United States*, chapter 1.

incomes during the twenties. These losses were in some cases intensified by mortgage debt incurred during the land boom. Thus, we turn our attention now to the high incidence of farm mortgage foreclosures in the 1920's.

B. Mortgage Foreclosures

The high rate of involuntary transfers of farm real estate after 1920 is another index often cited as proof of agricultural depression in the twenties. Between 1912 and 1920 less than five out of every thousand farms in the United States changed hands each year through bankruptcy, mortgage foreclosure, or assignment in lieu of foreclosure. After 1920, however, the number rose steadily until over eighteen farms per thousand in the nation were transferred involuntarily in 1927. The number dropped only slightly in 1928 and 1929, but rose sharply again after 1930 to a rate of almost 39 farms per thousand in 1933.[17]

17. Lawrence A. Jones and David Durand, Mortgage Lending Experience in Agriculture (NBER, Princeton, New Jersey, 1954), 6, Figure 2. Data taken from USDA.

Therefore, the high rate of involuntary real estate transfers after the early twenties not only seems to support the notion that farmers faced long-run economic disabilities after World War I, but it also suggests that agricultural conditions in the twenties were similar to those faced in the "great depression" of the early thirties.

Indeed, the tendency to equate farm foreclosure with economic depression was reinforced by events in the early thirties. There are numerous stories from that period of foreclosure proceedings where sheriffs were terrorized by angry mobs of farmers who were not willing to see a neighbor's home sold at auction. Furthermore, there is no doubt that the wave of farm foreclosures that affected much of rural America in the early thirties was due to countless farmers being unable to meet interest and principal payments in the face of falling farm prices and incomes. There is not enough evidence, however, to prove that farm foreclosures in the twenties were also caused by low prices and incomes. The connection that many writers draw between farm foreclosures and farm depression is usually circular. In other words, agricultural depression in the twenties is used to explain the high incidence of farm foreclosures and foreclosures in turn are cited as proof of agricultural depression.

We have seen the weaknesses in this kind of reasoning as it
has usually been applied to data on falling farm land values
in the twenties. What can be said about the same reasoning
when it is applied to data on farm mortgage foreclosures?

The type of published data that is available to study
farm mortgage foreclosures prevents meaningful analysis of the
causes of foreclosure in the 1920's, although this problem is
rarely recognized by historians and economists who have used
the data.[18] These data are inadequate mainly because the lowest
level at which involuntary transfer statistics are aggregated
is by individual states. State-wide data would be useful if
commodity, soil, climatic, and size-of-farm conditions in individual states were homogeneous; however, such homogeneity does
not exist at the state level. In fact, it does not exist at
the county level in most counties of the United States.
Therefore, even unpublished data on farm foreclosure rates
which the USDA has aggregated by "crop reporting district"

---

18. Jones and Durand recognized many of the weaknesses in published farm real estate transfer data and were able, to some
extent, to derive more meaningful statistics with county data
from an unpublished Works Progress Administration study completed
in 1939. See Ibid., 26-31.

do little to overcome the problem inherent in the published

state figures.[19] Some economists would no doubt contend that

farm foreclosure rates could be correlated with farm income

proxies such as farm land rents, also available from USDA on

a "crop reporting district" basis, to see if significant relation-

ships exist that prove or disprove the hypothesis that agricul-

tural depression caused these foreclosures. The explanatory

power of such correlations cannot be greater, however, than the

information contained in the data on which the correlations are

based. Therefore, the causes of farm mortgage foreclosures

in the twenties cannot be studied with any precision until

data comparing foreclosed and non-foreclosed farms are avail-

able either for individual farms or for samples of farms

where farm size, commodities, soil conditions and many other

factors are carefully identified and controlled. It is unlikely

that such data will be forthcoming until a great deal of res-

earch is done in county courthouse records across the United

States.

In the meantime, some limited evidence can be cited

that reveals contradictions in the usual notion that agricultural

depression in the twenties caused the rash of farm mortgage

---

19. The basic published source of farm real estate transfer
data by state is the bi-annual USDA bulletin The Farm Real Estate
Situation. For the twenties, these include Circular No. 150
(1925-29) and Circular No. 309 (1929-32). The raw data by "crop

foreclosures in that decade. For one thing, it seems that farm mortgage foreclosures in the hard-hit Midwest and Northwest regions were more concentrated geographically in the twenties (1926-1929) than in the early thirties (1930-1933).[20] If general agricultural price and income conditions had been the main factor causing foreclosures, one would expect the incidence of foreclosure to be pervasive. Although foreclosures do appear more widely dispersed in the early thirties, their greater concentration in the late twenties suggests that prices and incomes may not have been an important causal factor.

One reason why mortgage foreclosures were less pervasive in the twenties was the effect of climatic and crop-disease problems in certain regions, especially in the early twenties. We noted before that severe drought and wheat blight affected northern spring wheat areas for a long period after 1918. In

reporting district" (usually includes 15-20 counties) are available on worksheets from USDA.

20. Based on "crop reporting district" estimates of annual mortgage foreclosure rates from 1926, the earliest year data are available, for all districts included in states in the East North Central and West North Central census regions plus Montana, Idaho, Washington, and Oregon. The mean annual farm mortgage foreclosure rate in that area from 1926-1929 was 2.43 per cent of all farms, and in 1930-1933, 2.79 per cent of all farms. In the former period the standard deviation from the mean (1.95) divided by the mean (2.43) was .80, while the same ratio in the latter period was only .53, suggesting there was greater area-concentration of foreclosures in the earlier period.

fact, foreclosure rates in some North Dakota and Montana districts were among the highest in the nation in the late twenties.[21] A similar situation existed in the Southeastern cotton areas of Georgia and South Carolina where boll weevil infestation reduced yields drastically in the early twenties and led to high foreclosure rates later in the decade.[22] Nevertheless, some parts of the nation that escaped natural calamities still suffered high foreclosure rates in the twenties.

Any explanation for farm mortgage foreclosures in the 1920's must give high priority to the postwar land boom and to institutional factors that were involved in mortgage lending at that time. This seems to be a major conclusion of the most complete study of farm mortgage problems in the United States, published by the National Bureau of Economic Research in 1954.[23] Unfortunately most of the data compiled by the authors of this study lump the interwar years together as a unit instead of

---

21. See above, page 114. Also, L. Jones and D. Durand, Mortgage Lending Experience in Agriculture, 30 and 68-75.

22. Ibid., 189.

23. Ibid., 180-195. Jones and Durand's final statement (197-198) that prices and incomes were the main reason for interwar financial distress in agriculture is not supported by their study and seems to have been added-on as an afterthought.

considering the twenties and thirties as distinct periods.
Nevertheless, their analysis provides many useful insights
into conditions during the twenties. For instance, they found
that mortgage foreclosure rates after World War I were related
more to the staggering growth in farm mortgage debt itself
during the war decade than to the boom in land values on which
the mortgages were based.[24] This relationship is important
since it seems true that farm land values, at least in short-run
periods, rarely reflect potential trends in land income. Furthermore, it also seems that in a period of speculative activity
such as the farm land boom of 1919 and 1920, values rise relatively
more on lower priced and poorer quality soil than on more productive land.[25] Until very recently, it had been customary for
farm lenders in the United States to base mortgage loans on
the value of the mortgaged land rather than an estimate of its
future income earning power. Therefore, with evidence that poorer
quality land grew relatively more in value during the postwar
boom than better land, this lending practice presumably led to

24. Ibid., 182-185.

25. Ibid., 192-195. On the sluggish response of land values
to income, also see George Iden, "Farmland Values Reexplored,"
Agricultural Economics Research, USDA, Vol. XVI, No. 2 (April, 1964).

excessive loans on poor soil. In that case, one would expect
to find higher foreclosure rates in the twenties in poor soil.
regions, no matter what prices and incomes in general may have
been. This was found to be true in the NBER study, especially
in the Corn Belt region stretching from Ohio to eastern Nebraska.[26]

Another institutional factor that exacerbated the mortgage
distress problem of the twenties was the general type of credit
instrument used by lenders in that period.[27] Most farm mortgage
loans took the form of a one to five year note that was payable
or renewable (at the lender's option) at maturity. Long-term
amortizeable notes were not common until much later. Thus,
farmers normally expected to renew their notes from period to
period, as was common in the era of rapidly rising land values
from the later 1890's to 1920. If land values ever declined, however,
farmers were faced with refinancing in the short-run a type of
debt that could only be regarded as long-term. Except in such
cases, farm borrowers regarded the annual interest payments as
their only debt "burden." Thus, they rarely hesitated to "trade
on the equity" of their rising land values, especially in 1919
and 1920. In turn, bankers and other lenders often overextended

26. L. Jones and D. Durand, Mortgage Lending Experience in Agriculture, 87.

27. The material in this paragraph was drawn from Norman J. Silberling, The Dynamics of Business (New York, 1943), 145ff.

themselves under the illusion that short-term notes were highly liquid. It was inevitable that many loans extended under such conditions would become frozen as asset values dwindled after 1920.

An additional sign that farm mortgage foreclosures in the twenties resulted more from speculative and excessive borrowing in the immediate postwar period rather than from economic depression in agriculture, was the high proportion of foreclosures between 1920 and 1930 that were on junior rather that first mortgages.[28] These junior mortgages arose largely because few borrowers in the postwar boom period could produce anything but small down payments against the high land values at the time. In the extreme but all too common case, farm borrowers put down 10 per cent on an acre worth $200, borrowed 50 per cent on a first mortgage, and secured a second mortgage for the balance. Where a farmer merely borrowed on land he already owned, the second mortgage would be taken by a local bank and the first mortgage by an insurance company.

When land values fell after 1920, junior mortgage holders often risked losing their entire equity if the first mortgage

28. The discussion of junior mortgages in the two following paragraphs comes from William G. Murray, "Iowa Land Values: 1803–1967," The Palimpsest, Vol. XLVII, No. 10 (October, 1967), 464–465 and 469–472. Also a letter from Professor Murray to the author dated August 14, 1968.

holder foreclosed and auctioned the property. This risk was
greatest where the borrower was the farm operator and he
was finding it difficult to pay interest charges and taxes
based on wartime valuations. To prevent default and auction,
the junior mortgage holder in such cases would foreclose his
part of the debt and take over interest and tax payments.
In the meantime he would expect land values to recover enough
to liquidate the various mortgages at full value. Available
data do not permit even a guess as to how many farm mortgage
foreclosures recorded in the twenties were this type of
junior foreclosure. Nevertheless, one expert on agricultural
finance who witnessed the problem in the twenties first-hand
has referred to the foreclosure crisis of that era as the
"junior mortgage depression of the twenties."[29]

Therefore, the farm mortgage foreclosure problem in
the 1920's does not seem to bear any necessary relation to the
agricultural price and income situation during the decade itself.
Unlike the "great depression" period of the early thirties,
when there is little doubt that foreclosures resulted from

---

29. Ibid., 469.

low prices and incomes, farm foreclosures in the twenties

seem to reflect the aftermath of a speculative borrowing

spree during the postwar boom.[30] Many farm foreclosures in

the twenties were purely the result of bad luck and overopt-

imism, not the result of depressed farm incomes. More research

is needed, however, before precise estimates of the effect

of this optimism can be made. At this point it will be use-

ful to consider some aspects of the foreclosure problem that

were faced by one important class of lenders; namely, rural

banks.

C. Rural Bank Failures.

An unusually high number of bank failures occurred

in the United States during the 1920's. Data on bank suspen-

sions before 1892 are incomplete; however, it is likely that the

incidence of bank distress in the twenties was more severe

---

30. A good summary statement of the speculative nature of
the postwar land boom and its relation to later farm debt
problems is in L.J. Norton, "The Land Market and Farm Mortgage
Debts: 1917-1921," Journal of Farm Economics, Vol. 24, No. 1
(February, 1942).

and prolonged than in any prior period of the nation's history.
Over 5,700 banks suspended operations in the United States
between 1921 and 1929. The percentage of all active banks
that suspended in each of those nine years was higher than
in any single year since 1893, a year of severe economic
crisis in the nation.[31] In previous periods of American
history, however, bank suspensions were usually related to general
financial depression. The high incidence of suspensions in
the 1920's was unique in that most of the nation experienced
general prosperity during that decade.

Agricultural depression in the twenties has often
been cited as the main reason for this paradox of banking
crisis in the midst of general prosperity. The basis for this
assertion is the heavy concentration of bank suspensions and
failures in rural agricultural areas between 1921 and 1929.
Almost 47 per cent of all active banks in the United States in
1920 were in three major agricultural regions, West North Central,
South Atlantic, and Mountain; however, over 71 per cent of all
bank suspensions in the nation from 1921 to 1929 were in the
same three regions. Although 18.7 per cent of all active banks

31. Board of Governors of the Federal Reserve System, _Federal
Reserve Bulletin_, Vol. 23, No. 12 (December, 1937), 1204.

in the United States in 1920 suspended operations between 1921 and 1929, the percentages in the above three regions were 28.5, 23.7 and 33.7 respectively.[32] Furthermore, 91.6 per cent of the nation's banks suspensions from 1921 to 1929 occurred in towns of less than 10,000 population.[33] Presumably the economic life of these relatively small communities centered around agriculture, especially in the South and West.

One theory that linked these bank failures with farm depression was offered by Clark Warburton, research economist for the Federal Deposit Insurance Corporation in the late thirties. Warburton suggested that "...regional and local differences in the incidence of bank suspensions were probably associated with inter-regional balances of payments, and that these interregional balances of payments were associated with the relative prosperity or depressed conditions of the various industries of the nations." Assuming that agriculture was a

32. Board of Governors of the Federal Reserve System, _Federal Reserve Bulletin_, Vol. 23, No. 9 (September, 1937), 883. These regions conformed to Bureau of the Census definitions, except that Florida was not included in the South Atlantic region in these data. Florida's bank suspension rates in the 1920's, although extremely high, reflected almost entirely the impact of the mid-twenties Florida real estate debacle. These data exclude mutual savings and private banks.

33. Norman J. Wall and Lawrence A. Jones, "Short-Term Agricultural Loans of Commercial Banks, 1910-45," _Agricultural Finance Review_, USDA, Vol. 8 (November, 1945), 5.

depressed sector in the twenties, it followed that agricultural regions may have suffered payments outflows -- like nations with unfavorable trade balances. Presumably this outflow reduced bank deposits in agricultural regions, making it difficult to liquidate assets and thereby causing bank suspensions.[34]

The Federal Reserve System in 1937 expressed a similar idea when a study of bank suspensions in the 1920's concluded that:[35]

> "Agriculture, in particular, was passing through a period of readjustment incidental to the reduction of the prices of farm commodities and land after the war. The balance of payments of agricultural regions in the United States was unfavorable, and banks serving agricultural communities were under pressure. Suspensions among such banks were numerous throughout the 1921-1929 period."

The theories offered by Warburton and the Federal Reserve System were never supported by empirical evidence showing that

---

34. Clark Warburton, "Eleven Years of Research at the Federal Deposit Insurance Corporation," unpublished manuscript in FDIC library, Washington, D.C, (n.d.), 81-84.

35. Federal Reserve Bulletin (December, 1937), 1205.

agricultural regions in the United States had suffered payments
imbalances in the twenties. Nevertheless, the idea that agricultural depression caused the 1920's banking crisis has been
widely accepted.

On the other hand, many experts writing in the twenties
and thirties disagreed with that position and contended that
the banking crisis reflected internal deficiencies within
the banking system itself. Generally, those writers who explained bank suspensions in the twenties in terms of agricultural
distress were convinced that the banking system itself was
essentially sound.[36] One economic historian has suggested
that the prevalence of this attitude in the twenties "...led
to more complacence about bank failures than was later proved
to be justified when general depression overtook the country."[37]
In spite of this statement, however, it is clear that many
contemporary analysts of the bank crisis in the twenties placed
greater emphasis on structural flaws in the banking system
than on exogenous factors such as farm depression. The two

36. These differing viewpoints can be seen in testimony
before the U.S. Congress: House Banking and Currency Committee
(February 25, 1930 - July 11, 1930); Senate Committee on Banking
and Currency (January 19, 1931 - January 30, 1931).

37. G.H. Soule, Prosperity Decade, 152.

structural flaws most often cited were "overbanking," or too
many banks in the United States, and evidence of too many cases
of unsound bank management, especially in the period from about
1918 to 1920.

Conditions that favored "overbanking" were deeply rooted
in the American ideal of free enterprise, interpreted by many
states to mean that any man with the required capital should
be entitled to own and operate a bank. An effort had been
made to draw all banks into a uniform national system when
the National Banking Act of 1863 forced a tax on the notes issued
by non-national banks. The plan failed, however, because many
banks entered the field of deposit banking after the Civil
War. A tax on bank notes was not a prohibitive barrier to entry
in such cases. Therefore, state chartered banks blossomed in
the late nineteenth century as deposit institutions, formed
under state laws that often required minimum capitalization
as low as $10,000. As competition from these state banks grew,
however, protests from national bankers led to a reduction
in 1900 of the minimum capital required to form a national
bank, from $50,000 to $25,000. In the ensuing years of rising
prices and general prosperity, the number of all banks in the

nation grow from less than 14,000 institutions in 1900 to over 30,000 by 1920.[38] This surge in bank formation was especially prevalent in several midwestern states where many small institutions with incompetent managements flourished under lax banking laws -- a situation described by one authority as "too many banks with too few bankers."[39]

The notion of "overbanking" often rests on the assumption that an optimal relationship exists between the number of banks and size of population, beyond which excessive competition will develop for loans and deposits. Such competition supposedly generates high operating costs and excessive risk-taking so that margins become too narrow for profitable operations unless economic conditions are unusually favorable. In this light, the steady rise in the number of banks per capita from 1900 to 1920 suggests an appropriate prelude to the rash of bank suspensions after 1920. There was one bank for 5,560 persons in 1900, but by 1920 the ratio had risen to one for 3,515.[40] Furthermore, this growth of banks relative to the population was

---

38. C.D. Bremer, American Bank Failures (New York, 1935), 29. The material in this paragraph is outlined in chapter 1 of Bremer.

39. Clark Warburton, Deposit Insurance in Eight States: During the Period 1908-1930 , unpublished manuscript in FDIC library, Washington, D.C, (1959), "Nebraska," 49.

40. C.D. Bremer, American Bank Failures, 29.

not evenly distributed over the nation. In only three states
west of the Mississippi (California, Arkansas, and Arizona)
was the 1920 population to bank ratio higher than the national
average, while in only three states east of the Mississippi
(Vermont, Indiana, and Wisconsin) was the 1920 ratio lower
than the national average. Generally, the lowest population
to bank ratios were in predominantly agricultural states of the
Midwest and West.[41] Likewise, as the "overbanking" thesis
would predict, the average percentage of active banks suspended
between 1921 and 1929 was considerably higher than the national
average in most of those states. In one study, a correlation
of bank suspension rates from 1921 to 1929 with 1900 population
to bank ratios for all states of the United States revealed
a strong inverse relation between suspensions and population
per bank.[42] Therefore, depression in agriculture is not a
necessary condition to explain high bank suspension rates in
the 1920's. State banking laws that permitted "overbanking"
were perhaps a more important causal factor.

Studies of individual bank failures in several midwestern
states lend added weight to the notion that structural flaws

---

41. Ibid., 55.

42. Ibid., 56.

in the banking system contributed more than alleged depression

in agriculture to the high rate of suspensions in the 1920's.

These studies of individual cases, based largely on evidence

from bank examiner audit reports, revealed that a high per-

centage of failures resulted from unsound lending practices

and even fraud on the part of bank officials.[43] These practices

were especially pronounced in farming areas during the postwar

land boom. Many rural banks were caught after 1920 with

excessive amounts of mortgage credit extended from late 1918

to early 1920.[44]

Fred Garlock, an economist whose father was a banker

in Iowa during this period, outlined the probable factors

that led to a great deal of excessive bank lending during the

war in Iowa.[45] For one thing, bankers had been accustomed to

a predictable seasonal relationship between loans and deposits

in the decade before World War I. This relationship was upset

---

43. Clark Warburton, Deposit Insurance in Eight States, sections
entitled "causes of bank failures."

44. This would apply especially to banks that took on considerable
amounts of "junior" mortgages.

45. The material in the paragraph is from Fred L. Garlock,
"Bank Failures in Iowa," Journal of Land and Public Utility
Economics, Vol. II, No. 1 (January, 1926). In a discussion I once
had with Garolock he revealed that his father had purchased a
farm in Story County, Iowa for $100,000 in the last quarter
of 1919. The property was sold in 1946 for $30,000.

during the war as a secular increase in deposits left many banks
with excess loanable funds. With prices and land values rising
rapidly, bankers eagerly placed these funds with willing cus-
tomers. In fact, intense competition to place these funds
caused many bankers to relax normal credit standards. With the
rush of deposits in 1919, many loans that were placed could not
have been liquidated unless prices had continued to rise as
they had for the previous two or three years. Thus, most of
these loans became frozen assets when land values fell after
1920. Banks holding excessive amounts of loans made in 1918
and 1919 were ultimately forced to suspend operations.

Thus, there is a great deal of evidence to suggest that
the banking crisis of the twenties was caused more by "overbanking" and unsound lending, especially in rural areas, than
by agricultural depression. It seems that many, if not most
of the banks that suspended operations between 1921 and 1929
would have been forced to do so regardless what the economic
situation of farmers themselves may have been. Yet, it may be
too soon to lay most of the blame for bank suspensions in the
twenties on poor managmment and "overbanking." Our evidence
is more suggestive than exhaustive. Although "agricultural
depression" in itself does not appear to have caused the problem,

there is one additional factor that may have contributed to more
bank suspensions in the twenties than either "overbanking"
or unsound lending. This was the continuing shift of farmers'
trading activities from smaller to larger population centers,
which accelerated markedly after World War I under the influence
of better roads, autos, trucks, and telephones.

If most bank suspensions in the twenties had been due
to excessive competition from "overbanking," we would predict
that suspension rates varied among different size-classes of
banks. Presumably smaller banks would have been more vulnerable
than larger banks, because of higher unit costs and less ability
to attract competent management. In fact, smaller banks do
seem to have suffered suspension more often in the twenties.
Almost 65 per cent of all banks in the United States in 1920
had total loans and investments of less than $500,000; however,
85 per cent of all bank suspensions from 1921 to 1929 were in
that size category.[46] Nevertheless, this does not prove that
"overbanking" of small-sized banks was the main reason for the
suspensions. If the pressure of excessive competition on small

---

46. *Federal Reserve Bulletin* (December, 1937), 1216, Table
14 and 1217, Table 16.

banks had been the major problem, then suspension rates would
not necessarily have varied between different-sized towns.
If anything, one may expect that excessive competition would
have occurred more in large towns and cities where opportunities
to open new banks were more abundant. On the other hand, if
most bank suspensions resulted from a structural change in
the locus of rural trading activities and not from "overbanking,"
then suspension rates should have been higher in smaller towns
than in larger centers. It is noteworthy, therefore, that
while two-thirds of all banks in the United States in 1920
were in towns of less than 2,500 population, almost 80 per cent
of all bank suspensions from 1921 to 1929 were in those same
towns.[47]

Thus, it is likely that the high incidence of suspensions
among small banks in the twenties was due not so much to "overbanking" as it was to problems faced in small rural communities
where so many small banks were located. Furthermore, the

---

47. Federal Reserve Bulletin (September, 1937), 906, Table 11.
U.S. Federal Reserve System, Committee on Branch, Group, and
Chain Banking, Reports, unpublished mimeograph in FRS Library,
Washington, D.C. (10 vols., circa. 1932), Vol. 3, "Changes in
the Number and Size of Banks in the U.S., 1834-1931," 48, Table 21.
Russell A. Stevenson, ed., A Type Study of American Banking: Non-
Metropolitan Banks in Minnesota (Minneapolis, November, 1934), 16.

problems faced in these small rural communities do not seem

to have emanated primarily from economic distress among farmers.

Instead, the main "difficulty" appears to have been that farmers

were becoming increasingly mobile and were shifting their banking

and trading activities to larger urban centers. Almost no

research has been done to document the effect of the automobile,

truck, and highway improvements on rural America after World

War I.[48] Two very limited studies published by the U.S. Depart-

ment of agriculture in the late twenties revealed that up to

one-third of the farmers in certain areas who purchased trucks

in the mid-1920's changed their major trading center to

one that usually was twice as far from home as the old center.[49]

---

48. Numerous writers have made passing remarks about the impact of autos, trucks, mail-order business, and telephones on rural business after 1920. See works cited above by C.D. Bremer, Clark Warburton, Russell Stevenson, as well as comments to the Senate Committee on Banking and Currency (January 19 - January 30, 1931) by J.W. Pole, Comptroller of the Currency, and George L. Harrison, Governor of the Federal Reserve Bank of New York. Also see Charles W. Collins, Rural Banking Reform (New York, 1931), 66, and H. W. Peck, "The Influence of Agricultural Machinery and the Automobile on Farming Operations," The Quarterly Journal of Economics, Vol. XLI (May, 1927), 534-544.

49. H.R. Tolley and L.M. Church, Motor Trucks on Corn Belt Farms, USDA, Farmers' Bulletin No. 1314 (Washington, D.C., March, 1923). L.M. Church, Farm Motor Truck Operation in the New England and Central Atlantic States, USDA Department Bulletin No. 1254 (September 27, 1924).

Nevertheless, such shifts in farmers' trading patterns
may have contributed to the demise of many small-town banks
that were otherwise overextended by the early twenties. The
small rural bank could not offer the low-cost range of services
available at banks in larger towns. It is notworthy that much
of the force behind the McNary-Haugen movement came from rural
bankers. One writer noted in 1926 that the "Executive Committee
of 22," a group of Corn Belt spokesmen who lobbied in behalf
of the McNary-Haugen cause, developed from a conference of the
Iowa State Banker's Association.[50] Historians writing about
agricultural depression in the 1920's have ignored the idea
that much of the "farm" protest in that decade perhaps reflected
the economic distress of rural bankers and merchants rather
than of farmers themselves. Our discusssion here suggests
that the banking crisis itself was rooted in overindulgence in mortgage lending during the postwar land boom and in structural shifts
in rural economic activity that were caused by growing mobility
of farmers. The alleged depression in agriculture does not
appear to have caused the rash of bank suspensions in the
decade after World War I.

----

50. Eric Englund, "The Bank's Part in the Farmer's Trouble,"
Nation's Business, Vol. XIV, No. 11 (October, 1926), 50.

The material in this chapter suggests that "agricultural depression" does not explain the adverse trends in farm land values, farm mortgage foreclosures, or bank failures in the 1920's. Instead, the trends in these indices of distress appear to have flowed from a wave of optimism about farm prices and land values that swept over rural America in the immediate postwar period. Wartime optimism cannot explain, of course, all of the land, mortgage, and bank problems faced in farm areas during the 1920's; however, it seems to have been the single most important factor. Precise judgement on the matter must wait until a great deal of research is done in previously untapped sources. The highly aggregated quantitiative data that historians and economists have used so far to study these problems can give only limited, often misleading, answers.

A very important field of research suggested by the material in this chapter is to study the shifting locus of economic activity in small rural towns in the United States following the advent of the auto and truck. Most of the agitation for farm relief in the twenties came from rural areas; however, the problems faced by many who cried the loudest may have been quite apart from any problems faced by farmers themselves. The discussion of bank failures in the 1920's suggests there may

be a great difference between rural and agricultural economic

distress.

Nevertheless, many farmers surely faced economic problems

in the period after World War I. We have shown, however,

that the conventional explanations for farmers' grievances

after 1920 are not supported by evidence on farm incomes, prices,

and production trends in that period. What remains is to locate

some of the factors that did pose a challenge for farmers in

the twenties, and to assess the significance of these factors.

Accordingly, the next chapter will review the major trends in

farm output and inputs from 1900 to 1930 to see what conditions,

if any, were significantly different in the twenties.

CHAPTER VI

AGRICULTURAL PRODUCTION IN THE TWENTIES:
CONTINUITY AND CHANGE

Our criticism of the notion of agricultural "depression"
in the 1920's has focused primarily on the statistical artifacts
that scholars use to depict economic conditions in American
farming after World War I. The discussion in previous chapters
suggests that data on parity prices, surpluses, land values,
mortgage foreclosures, and bank failures reveal a confusing
and ambiguous picture, especially when viewed in a regional perspective. It seems that economic distress in agriculture during
the twenties rarely, if ever, resulted from overproduction
and low prices -- the causes cited so often by economists and
historians. In certain regions of the country, crop failures
due to adverse climatic and biological conditions apparently
caused the difficulties faced by many farmers. Furthermore,
it is not even clear that the "farm" crisis of the twenties was
primarily an agricultural phenomenon. Much of the economic
distress usually attributed to farm problems in that decade
may in fact have reflected rural non-farm problems. In any

event, the kinds of quantitative data used by agricultural historians and economists make it almost impossible to picture what actually happened on American farms after World War I.

This chapter will briefly sketch the major trends in commodity production and resource employment in American agriculture from 1900 to 1930. The discussion will draw heavily on data from published census material. Most of these data have been available for over thirty years; however, historians have concentrated their attention so much on the political aspects of agricultural history in the twenties that economic aspects outlined in contemporary political tracts have been taken for granted. Almost no quantitative data beyond the conventional price and production series have ever been developed or analysed. This chapter presents regional data on trends in farm output and inputs from 1900 to 1930 which, for the most part, are not available elsewhere. Although the entire thirty year period after 1900 will be covered, changes that occurred after 1919 will be stressed. The twenties will thereby be viewed in the perspective of the long-run, usually ignored by most writers, to discover if agricultural trends after World War I reflected continuity or change from conditions in earlier decades.

A. Farm Output, 1900-1930.

An important feature of twentieth century American agriculture is that production of most commodities is concentrated
in specific regions. Distinct wheat, corn, dairy, cotton and
truck farm regions, less apparent in earlier periods of rapid
westward expansion, were clearly discernible by 1900. This
feature must always be remembered since it makes it almost
impossible to generalize about agricultural conditions in the
nation as a whole. Economic historians, policy-makers and others
too often consider American agriculture as an industry in itself
rather than a conglomeration of industries, each facing different
demand and price conditions at the same time. For instance,
farmers in different regions will respond quite independently
if at the same time the demand for fluid milk is strong while
the demand for wheat is weak. Attempts to study price and
demand trends in the aggregate or to study the conditions
faced by one commodity as if it were representative of agricultural conditions in general inevitably produce a distorted
picture.

The pattern of commodity specialization in the United
States can be studied by calculating the percentage share of
production of various commodities in each of the ten farm
production regions as they are defined by the U.S. Department
of Agriculture. Tables VI - 1 through VI - 4 show regional
distributions of the production of all farm products as well
as several important commodities at decade intervals from 1899
to 1929.[1] In order to study <u>changes</u> in the patterns of regional
specialization, these tables show the standard deviation from
the mean percentage of each regional distribution at each
decade. With ten farm production regions, the standard deviation
for any year would be zero if each region produced 10 per cent
of the national output of any given commodity. Likewise, the
standard deviation will be larger in magnitude the more production is concentrated in some regions rather than in others.
Therefore, if the standard deviation of the distribution for some
commodity increases from one period to another, we may infer

---

1. See the map on page 209 which outlines the ten USDA farm
production regions. The percentages in these four tables are
drawn from census data which must of course be used with caution.
The usual pitfalls of under-or over-enumeration and classification
changes that one encounters using census data are minimized here,
however, where trends in relative rather than absolute magnitudes
are being studied.

that production of that commodity was more concentrated regionally at the latter date. On the other hand, a decrease in the standard deviation may reflect increased dispersion in the locus of production.[2]

Turning to Table VI - 1, the percentage of the value of all farm products used, traded or sold in each production region at each census year is a crude gauge of the significance of various regions as agricultural producing areas. The declining standard deviations suggest that agricultural production became more widely distributed among the various regions of the nation between 1899 and 1929. Within the nation, the regions west of the Mississippi River gained in relative importance; however, the decline in eastern regions was confined almost entirely to the Northeast and Corn Belt states where an increasing share of the labor force was shifting into the industrial and service sectors. The steady relative growth of agricultural production in the Plains and Mountain regions reflects the continuing importance in the twentieth century of new acreage opened in frontier areas. The Pacific region exper-

2. There are several ways to gauge relative concentration in statistical distributions, of which the standard deviation from the mean is but one. Another measure of concentration, thought by some to be less ambiguous than the standard deviation, is the Gini coefficient which is derived from Lorenz-curve data.

TABLE VI - 1

REGIONAL SHARES IN THE VALUE OF ALL FARM PRODUCTS SOLD
OR USED IN THE UNITED STATES BY CENSUS YEARS

1899 - 1929

	1899	1909	1919	1929
United States	100.0 %	100.0%	100.0%	100.0%
Northeast	15.2	12.0	10.5	11.1
Appalachian	9.5	9.5	10.3	9.2
Southeast	6.0	7.4	7.5	6.2
Lake States	9.8	9.4	9.9	9.9
Corn Belt	29.4	27.6	25.3	20.6
Delta States	5.4	4.9	5.0	6.0
Northern Plains	10.6	12.9	11.4	12.2
Southern Plains	6.6	7.6	9.7	9.6
Mountain	2.8	3.8	4.3	5.2
Pacific	4.3	4.8	6.0	9.2
standard deviation	7.3	6.5	5.6	4.2

Sources: Calculated from state totals found in the following:
1899-Twelfth Census of the U.S., 1900, Agriculture, Part I,
cxxii-xcciii; 1909 and 1919- Fourteenth Census of the U.S.,
1920, Agriculture, Volume V, 18; 1929-Sixteenth Census of the
U.S., 1940, Agriculture, Volume III, 905-911.

ienced the largest percentage increase in importance as an
agricultural area in this period. An important factor there
was rapid urban growth, especially in California in the 1920's.
Probably more crucial was the growing demand for specialty
crops in a wide national market.

The types of commodities produced differ markedly between
regions. Therefore, shifts over time in the relative importance
of various commodities can affect the contribution made by
any one region to the value of the nation's total agricultural
output. Five basic commodity groups that comprised about 65
to 70 per cent of the value of farm output in the United
States from 1899 to 1929 were wheat, dairy products, cotton,
hogs, and cattle.[3] The percentage changes in net output for
each of these groups from 1898/1900 to 1928/1930 were as follows:
dairy products, 69; cotton, 41; hogs, 37; wheat 30; cattle, -3.2.
Likewise, farm prices increased over the same period by the
following percentages: dairy products, 182; cotton, 164; hogs,
158; cattle, 147; wheat, 69.[4] Withal, the value of wheat produced

For an excellent discussion of concentration indexes see Eugene
M. Singer, Antitrust Economics: Selected Legal Cases and Economic
Models (Englewood Cliffs, New Jersey, 1968), 136-155.

3. See Table VI - 5.

4. Output and price data from the relevant tables in Frederick
Strauss and Louis H. Bean, Gross Farm Income and Indices of Farm
Production and Prices in the United States: 1869-1937, USDA, Tech-
nical Bulletin No. 703 (Washington, D.C., December, 1940).

in the nation over those three decades grew the least while the
value of dairy products, followed by cotton, hogs, and cattle
grew the most. Therefore, regions specializing in wheat
production during that period would have had greater difficulty
maintaining the same share of the value of all farm output in
the nation than would regions specializing in dairy, cotton,
or livestock production.

Looking within the separate commodity groups, the
regional distribution of dairy production is shown in Table VI - 2.
The enormous growth in the amount of whole milk sold from
American farms in the 1920's is striking indeed. It is especially
striking in view of the very small change in the number of
milk cows between 1919 and 1929 and the lack of evidence in
the literature that the productivity of milk cows grew very
much over the period. Note, however, that the census estimates
for whole milk include only the portion sold from the farm and
exclude the portion consumed by the farm family or fed to
livestock. As the distribution of milk cows suggests, this exclusion in the census understates total milk production in southern
and plains regions where access to urban markets was far more
limited than in eastern, central and western regions. The regional
distribution data reveal that farm sales of whole milk grew

TABLE VI - 2

REGIONAL SHARES IN THE NUMBER OF MILK COWS ON FARMS AND IN SALES OF WHOLE MILK AND BUTTER FROM FARMS IN THE UNITED STATES AT CENSUS YEARS, 1899 - 1929

	Milk Cows on Farms				Whole Milk Sold from Farms				Butter Sold from Farms			
	1899	1909	1919	1929	1899	1909	1919	1929	1899	1909	1919	1929
United States	100.0%	100.0%	100.0%	100.0%	100.0%	100.0%	100.0%	100.0%	100.0%	100.0%	100.0%	100.0%
Northeast	21.5	18.1	18.0	14.2	42.4	48.9	44.4	33.7	29.5	22.7	29.4	24.7
Appalachian	9.1	9.3	9.4	9.0	2.2	2.7	3.3	5.0	4.7	9.2	17.0	20.1
Southeast	4.5	5.2	5.3	4.1	.4	.5	.8	1.3	1.3	2.2	4.9	10.4
Lake States	13.5	15.9	19.4	19.4	19.3	21.9	25.6	26.1	16.1	15.6	8.4	5.0
Corn Belt	26.8	23.6	20.4	21.5	25.6	18.6	15.5	16.6	28.7	29.2	20.6	14.7
Delta States	4.7	5.4	5.2	4.8	.5	.5	.5	1.3	1.4	1.7	2.7	4.0
Northern Plains	9.3	9.7	8.7	10.8	4.4	1.1	1.5	2.3	10.2	9.1	5.3	4.4
Southern Plains	6.6	7.5	6.6	8.1	.5	.7	1.1	2.4	2.2	4.7	6.1	13.8
Mountain	1.9	2.5	3.2	3.9	1.5	1.6	1.9	3.0	1.4	1.8	3.5	2.1
Pacific	3.1	4.0	4.7	5.0	3.8	4.4	6.5	9.9	4.5	3.8	2.2	.7
standard deviation	7.8	6.6	6.4	6.1	13.6	15.0	13.8	11.0	10.6	9.1	8.7	7.8
United States: quantity	17.1	17.1 million	19.7	20.5	18.4	16.7 billion pounds	21.8	38.3	518	415 million pounds	208	135

Source: Calculated from state totals in U.S. Department of Commerce, U.S. Census of Agriculture: 1950, Volume II: milk cows, 403; whole milk, 410; butter sold, 411.

relatively far more in these less urbanized areas during the
twenties than in the highly urbanized central and eastern
regions. Therefore, the sharp growth in national whole milk
sales in the 1920's does not seem to reflect an increase in
total milk production as much as an increase in the percentage
of total milk production that was being sold, especially in
southern and plains regions. The standard deviations suggest
that income from farm sales of fluid milk became more widely
dispersed over the United States in the twenties. Furthermore,
a large part of the increased sales of fluid milk in that decade
apparently did not require equivalent investments in dairy
animals. The milk cows were already there before 1919.
The investments that made the enormous growth in whole milk sales
after 1919 possible were probably in trucks and road improvements,
giving easier access to urban markets in areas where farmers
previously marketed most of their milk output in the form
of butter.

Livestock production figures are not available on a
regional basis for this period, even in the census. The figures
in Table VI - 3 show regional shares in the nation's inventory of
livestock and suggest the importance of meat animals in each
region. Throughout these thirty years, the Corn Belt region

TABLE VI - 3

REGIONAL SHARES IN THE NUMBER OF CATTLE AND CALVES (EXCLUDING MILK COWS) AND HOGS AND PIGS IN THE UNITED STATES AT CENSUS YEARS, 1899 - 1929

	Cattle and Calves (excluding milk cows)				Hogs and Pigs			
	1899	1909	1919	1929	1899	1909	1919	1929
United States	100.0%	100.0%	100.0%	100.0%	100.0%	100.0%	100.0%	100.0%
Northeast	6.0	6.3	4.2	6.1	4.3	4.4	4.5	2.6
Appalachian	5.6	6.4	8.0	5.9	11.0	9.5	10.4	7.1
Southeast	4.0	5.3	2.8	3.3	6.3	7.8	8.7	5.7
Lake States	6.4	7.2	8.2	9.8	7.4	7.9	8.6	9.8
Corn Belt	21.1	19.3	20.4	18.6	43.1	40.2	39.2	43.9
Delta States	3.2	4.3	4.5	3.6	6.0	7.1	6.1	4.0
Northern Plains	16.4	14.8	17.2	17.5	13.9	13.4	12.8	18.5
Southern Plains	22.7	17.0	14.8	16.2	6.2	7.2	6.0	4.6
Mountain	11.0	12.6	14.9	13.6	.6	1.1	2.0	2.2
Pacific	4.0	5.7	5.3	5.8	1.7	2.1	2.4	1.9
standard deviation	7.0	5.2	6.0	5.6	11.6	10.7	10.3	12.2
		million				million		
United States: quantity	50.6	44.7	47.0	43.4	62.9	58.2	59.3	56.3

Source: Calculated from state totals in U.S. Department of Commerce, U.S. Census of Agriculture:

maintained a dominant position in stocks of cattle and hogs, although the relative importance of the Northern Plains region in stocks of hogs grew sharply, especially after 1919. At the same time the relative importance of hogs declined in southern regions. Comparing 1899 with 1929, cattle inventories were more widely dispersed among regions at the latter date, while hog inventories had become more concentrated.

Regional changes in the production of the nation's two most important cash crops are detailed in Table VI - 4. Cotton production was somewhat more widely dispersed in 1929 than in 1899, although there were no striking regional shifts in the locus of production until the latter part of the second decade. By 1919 and into the twenties, the effect of the boll weevil and the difficulty of competing with lower production costs in western areas caused a sharp drop in the importance of the South-east region as a cotton producer.[5] In wheat, the most significant change between 1899 and 1929 was the pronounced and steady growth in importance of the Plains and Mountain regions which led to the greater concentration of national production noted by 1929. The only marked deviation from this trend occurred in 1919 when the share of production in the Corn Belt region rose. As noted

5. Carter Goodrich, Migration and Economic Opportunity (Philadelphia, 1936), 141 and 149.

TABLE VI - 4

REGIONAL SHARES IN UNITED STATES WHEAT AND COTTON PRODUCTION BY CENSUS YEARS, 1899 - 1929

	Wheat (threshed)				Cotton			
	1899	1909	1919	1929	1899	1909	1919	1929
United States	100.0%	100.0%	100.0%	100.0%	100.0%	100.0%	100.0%	100.0%
Northeast	6.8	6.0	4.8	4.2	-	-	-	-
Appalachian	6.6	4.4	3.9	2.3	7.4	8.5	10.5	9.6
Southeast	-	-	-	-	35.8	42.5	34.2	23.3
Lake States	18.9	11.1	6.9	4.4	-	-	-	-
Corn Belt	22.9	20.6	27.6	13.6	-	-	1.0	1.4
Delta States	-	-	-	-	23.4	20.8	19.3	28.1
Northern Plains	25.1	42.3	31.6	41.4	-	-	-	-
Southern Plains	5.0	2.4	10.8	11.9	29.5	28.3	35.1	33.6
Mountain	2.8	4.2	5.6	12.0	-	-	-	1.6
Pacific	11.0	8.7	8.3	9.4	-	-	-	1.7
standard deviation	8.8	12.2	10.3	11.5	14.1	14.5	13.7	12.6
		million bushels				million sq. bales		
United States	659	683	945	801	9.5	10.6	11.4	14.6

Source: Calculated from state totals in U.S. Department of Commerce, U.S. Census of Agriculture: 1950, Volume II: wheat, 558–559; cotton, 570–571.

before a significant amount of corn acreage was diverted to wheat production under the influence of relatively more favorable prices for wheat during the war. At the same time the share of production in the Northern Plains and Mountain regions was reduced in 1919 as a result of severe drought in the spring wheat belts.[6]

These comments on the changing pattern of commodity specialization among regions provide a few useful insights into the problems of American farmers after World War I. For one thing, wheat production was becoming more specialized so that the trend toward greater regional concentration is not surprising. Although wheat prices did not rise as much as the prices of the four other commodity groups we studied, wheat production was becoming centered more and more in the Southern Plains and Mountain regions where conditions for wheat-growing were very favorable. On the other hand, fluid milk sales were becoming more widely dispersed among regions. This increased dispersion is significant since fluid milk demand and prices grew substantially throughout this thirty year period. It is necessary, however, to turn from this cursory review of

6. See above, chapter IV.

regional production distribution to a more detailed review
of changes in the composition of national agricultural output
from about 1900 to 1930.

A picture of the changing composition of the total value
of net farm output in the United States during the three
decades before 1930 is shown in Table VI - 5. The percentages in
this table are based on the estimated amount, valued at farm
prices, of output produced either for sale to non-farm pruchasers
or for farm family consumption, excluding the output of crops
used on the farm for seed or feed and milk fed to livestock. Feed
or seed consumed on the farm where produced or sold from one
farmer to another are excluded from these data to avoid counting
inputs twice, once as a raw material and again as finished crops
or livestock. Although this exclusion is necessary in gross income
accounting, it makes the resulting data less useful where changes
in total production are to be studied; however, more complete
data are not readily available and it is not difficult to assess
the bias between trends in net and gross output figures. Furthermore, these data include the value of changes in farm inventories of livestock in order to make the value data for livestock
comparable to that   for crops. The crop figures automatically

TABLE VI – 5

PERCENTAGE COMPONENTS OF THE TOTAL VALUE OF NET FARM OUTPUT IN THE

UNITED STATES AT DECADE INTERVALS

CROPS	1898–1900	1908–1910	1918–1920	1928–1930
Food grains	11.8%	10.7%	12.4%	7.4%
Wheat	10.5%	8.9%	10.2%	5.9%
Rye	.3	.3	.6	.2
Barley	.5	.8	.6	.4
Buckwheat	.1	.1	.1	–
Rice	.2	.3	.5	.3
Dry beans	.2	.3	.4	.6
Feed crops	9.6	10.0	7.7	4.9
Corn	5.4	5.7	4.9	3.7
Oats	1.8	2.2	1.7	.9
Hay	2.4	2.1	1.1	.3
Cotton	9.4	11.2	11.0	10.1
Fruits	3.3	3.1	3.7	4.5
Orchard	2.7	2.4	2.5	2.7
Citrus	.2	.4	.6	1.3
Grapes	.4	.3	.6	.5
Potatoes	3.2	3.7	3.8	3.3
White	2.7	3.1	3.1	2.7
Sweet	.5	.6	.7	.6
Oil crops	1.5	2.0	1.8	1.7
Flaxseed	.5	.4	.2	.3
Cottonseed	.7	1.2	1.4	1.2
Peanuts	.3	.4	.2	.2
Sugar	.4	.5	.6	.6
Cane	.3	.2	.1	.1
Beet	.1	.3	.5	.5
Tobacco	1.6	1.5	2.5	2.3
Total crops	40.5	42.7	43.5	34.8

(continued on the next page)

TABLE VI - 5 (continued)

PERCENTAGE COMPONENTS OF THE TOTAL VALUE OF NET FARM OUTPUT IN THE

UNITED STATES AT DECADE INTERVALS

	1898-1900	1908-1910	1918-1920	1928-1930
ANIMAL PRODUCTS				
Hogs	12.3%	13.0%	14.2%	12.5%
Beef	14.0%	9.4	8.9	10.9
Cattle	13.0%	8.4%	7.8%	9.4%
Calves	1.0	1.0	1.1	1.5
Dairy products	11.6	12.5	11.5	16.9
Manufactured	7.2	7.8	6.6	9.4
Fluid milk	4.4	4.7	4.9	7.5
Poultry products	7.9	9.3	9.4	12.2
Chickens	2.9	3.0	3.0	4.6
Eggs	5.0	6.3	6.4	7.6
Sheep and lambs	1.3	1.0	.9	1.3
Wool	1.0	1.0	.9	.9
Total animal products	48.1	46.2	45.8	54.7
TOTAL CROPS (prior page)	40.5	42.7	43.5	34.8
OMITTED PRODUCTS	11.4	11.1	10.7	10.5
Total all products	100.0%	100.0%	100.0%	100.0%

Source: Compiled from calendar-year gross income data in F. Strauss and L.H. Bean, Gross Farm Income and Indices of Farm Production and Prices in the United States: 1869-1937 (1940).

include inventory changes since they are computed from data on

crops <u>harvested</u>. Considering the comprehensive detail in the

study from which these data are drawn, these figures give

as accurate a picture of changes in farm output as is necessary

here.[7]

In general, the changes shown in Table VI - 5 reflect

the growth of real per capita disposable income in the United

States during the period involved, as well as differences in

the income elasticity of demand for various food products.

Other factors such as changes in export demands and the incidence

of war upset this generalization, but for the most part changes

in the level of per capita disposable income in the United States

have determined demand conditions faced by the American farmer.

Real disposable income per capita rose by about 70 per cent

from 1899 to 1929, with the growth from 1919 to 1929 alone

being about 20 per cent.[8] The proportion by which an individual

changes his spending for food as his income rises will vary

from one type of food to another. Basically, a far greater

percentage increase will occur in the spending for such things

as fruit, dairy products, poultry, and meat than for staples

7. The percentages in Table VI - 5 were compiled from Bureau of
Agricultural Economics data which one might imagine had been studied
as thoroughly as possible in the Strauss and Bean study. Nevertheless, Strauss and Bean were concerned with trends over a much
longer period than I am studying here. Consequently, I have had
to compile their data at decade intervals in order to study
shorter time periods.

8. Percentages compiled from U.S. Department of Commerce,

such as bread and potatoes.

The sharp decline in the share of wheat in the value of net farm output and the rising importance of fluid milk are therefore not surprising. The contribution to the total value of net farm output from hog production remained fairly stable over the thirty year period. The increase shown for the postwar years reflected heavy European demand for lard and pork. The decline in beef, although marked, occurred entirely in the first decade of the century, and is related to a sharp drop in the share of total beef production entering the export market. As the number of cattle per capita in the United States declined in the early 1900's and as Argentina rose as a major supplier of beef to world markets, American livestock farmers never again regained the position they held in the export beef market in the late nineteenth century.[9] The growing importance of fruit, especially in the citrus category, requires little comment. The percentage contribution of citrus fruits to the total value of net farm output more than doubled during the twenties as Florida and the lower Rio Grande Valley of Texas added to a

Historical Statistics of the United States: current disposable income, 139, F-9; implicit price index (GNP deflator), 139, F-5; population estimates, 7, A-2.

9. H. Barger and H.H. Landsberg, American Agriculture, 1899-1939, 101ff.

growing volume of production once centered almost exclusively

in California. Finally, the importance of tobacco rose, es-

pecially during the war, as the growing popularity of cigarette

smoking provided an expanding market for tobacco growers.[10]

The most noticeable change in the composition of the value

of net farm output between 1899 and 1929 is the shift in import-

ance from crops to animal products. To some extent the significance

of this shift is exaggerated by the nature of the data, since

the figures for feed crops reflect only the portion of such

crops sold off the farm. Total production of feed crops

actually rose over the period, although the percentage going into

commercial sales declined. Thus the shift from crops to animal

products shown in the table reveals that changing relative prices

during these years made it more attractive for farmers to process

food crops and sell them in the form of animal products. For

example, note that between about 1899 and 1929 the average

farm price of hay in the United States rose by a little over 50

per cent while the average farm price for dairy products rose

almost three times.[11] Thus, while 20.5 per cent of the total

10. Ibid., 71

11. F. Strauss and L.H. Bean, Gross Farm Income and Indices of
Farm Production and Prices, 62 and 98.

United States hay crop in 1898/1900 was sold to feed animals in

cities, by 1928/1930 this ratio had fallen to less than 4.1

per cent as the growing use of auto and truck transportation

led to the demise of the city horse.[12] At the same time, total

production of tame hay, which made up about 90 per cent of all

hay produced in 1929, rose from about 54 million tons in 1899

to about 76 million tons in both 1919 and 1929.[13] In spite

of the declining importance of hay in the total value of net

farm output, total production grew steadily after 1900 and the crop

remained important as livestock feed -- even during the twenties

when the tractor first began to displace horses on farms.

Looking more closely at the data in Table VI - 5, it is

significant that the shift from crop to animal products production,

both in the aggregate and within the sub-categories shown, did

not take place until the decade of the twenties. This is especially

apparent when comparing the trend in feed crops with that for

the dairy and poultry categories. The decline in relative import-

ance of feed crops in 1918/1920 was due to more favorable

prices for food grains during the war period. Note that the

---

12. Ibid., 62, Table 24.

13. Ibid., 59

contribution of all grains to the total value of net output
remained about the same from 1898/1900 to 1918/1920. The
divergence from trend at the latter date in the food and feed
grain sub-categories merely reflects a trade-off in which attention was focused on the food grains, so much in demand in Europe
at war's end, and less emphasis was put on feed, livestock
and dairy production -- commodity areas that were temporarily
priced out of the picture. If there had been no war, food grains
probably would have made up about 9.5 per cent of gross income
in 1918/1920, feed grains about 8 to 9 per cent, and the meat
(except hogs), dairy, and poultry areas would have contributed
a somewhat larger share than that shown. Withal, the rate of
change during the twenties still would have been the significant
feature in the table. Notice that for hay, a commodity affected
relatively much less by the war, the rate of decline from 1918/1920
to 1928/1930, was almost 60 per cent higher than the rate of
decline in the decade prior to 1918/1920. There is little question
that the most startling shifts from 1899 to 1929 in the composition
of commodities making up the nation's total farm output occurred
in the decade of the twenties.

In summary, it is clear that major changes occurred

in the composition of United States farm output during the 1920's.
Likewise, important changes occurred in the structure of farm
output in the various regions of the nation. In both cases,
the changes discussed here were underway in the early part
of the century; however, we found that certain major trends
accelerated after 1919. This was true of the pronounced shift
in the importance of fluid milk sales in less urbanized regions
after the war. Likewise, the trend away from crop sales to
sales of animal products accelerated markedly in the twenties.
Most of these trends and the accelerated rate at which they
occurred after the war were a result of basic changes in technology and demand in the society at large. The farmer himself
had little influence over these factors; however, he had to adjust
to the resulting changes in market demand or suffer economic
loss.

Some adjustments were relatively easy to make and provided
the farmer with added income. A case in point would have been
the farmer who found it easier to market fluid milk rather than
butter once trucks and surfaced roads became a reality. Other
adjustments were more difficult, as for instance the problem
faced by many farmers in upstate New York who had been accustomed
to supplying hay for the New York City market before World War I.

As the auto and truck replaced city horses, such farmers had the
choice of developing other markets, shifting into other commodities such as dairy products, or perhaps leaving farming
altogether. It is obvious that many farm operators adjusted
successfully to changing market demand in the twenties. On the
other hand, many probably failed. If the latter group constituted
a sizeable proportion of the troubled farmers who experienced
agricultural "depression" in the twenties, historians have
remained silent on the matter. One could hardly say that such
farm failures resulted from wartime expansion, overproduction or
low prices. Future research on agricultural problems in the 1920's
must consider the sub-group of American farmers whose problems
stemmed from failure or inability to adjust to changes in market
demand. To some extent, these adjustments required changes in the
types of production inputs used by individual farmers. Therefore,
we turn now to a review of the basic changes in the structure
of inputs in American agriculture from 1900 to 1930.

B. Farm Inputs, 1900 - 1930

The significance of the marked shifts in the composition

of agricultural output during the twenties is no less important
than changes in the pattern of inputs used to produce this
output. One only has to consider the commonly known accounts
of the adoption of tractors and tractor-powered implements
after 1919 to recognize this fact. The American farming scene
during World War I was still dominated by the combined power of
horse and man. Many competent analysts at the time were dubious
about the usefulness of the internal combustion engine in farming.[14]
Nevertheless, there was no doubt about acceptance of machine
power by the end of the twenties. Wheras an estimated 85,000
tractors were used on American farms in 1918, by 1929 the number
had grown to 827,000.[15] This widespread adoption of tractors,
as well as changes in other inputs, marks the 1920's as unique
in the thirty year period we are studying here. Farmers in all
regions, whether affected by these changes directly or not,
found their decisions and their economic welfare influenced by
the new technology.

14. H.R. Tolley, "The Farm Power Problem," Journal of Farm Economic
Vol. III, No. 2 (April, 1921). The War Industries Board did not
regard agricultural machinery production as essential to the
war effort since the labor saved by machines on the farm was
thought to be less than the labor consumed in machine production.
Also, The Percheron Society of America conducted a vigorous
campaign against installment financing of tractors, an item they
considered a wasteful investment. See Arthur G. Peterson, "Gov-
vernmental Policy Relating to Farm Machinery in World War I,"
Agricultural History, Vol. 17, No. 1 (January, 1943).

15. Martin R. Cooper, Glen T. Barton, and Albert P. Brodell,
Progress of Farm Mechanization, USDA, Misc. Publication 630

The following pages will consider trends for broad classes
of production inputs such as land, labor and capital. Only
limited remarks will be made about changes within each of these
input classes. Most of the data used here are compiled by farm
production region at decade intervals from 1900 to 1930. Emphasis
will be placed on the decade of the twenties, as in our previous
discussion of farm output.

The most general index of all production inputs in agriculture, prepared by USDA, shows an increase from 73 in 1900 to
98 in 1929 (1947-1949 = 100).[16] This index combines all inputs,
although it is more useful to analyze separately the trends in
inputs purchased from the nonfarm sector and inputs derived
from the farm sector itself. Charles Meiburg of the Food
Research Institute has prepared separate series for nonfarm
inputs (including processed feed and seed, fertilizer, motor fuel,
irrigation aids, veterinary services, and other items charged to
current expense) and net farm inputs (including farm labor,
real estate, and durable capital such as tractors and machinery).

(Washington, D.C., October, 1947) Table 41, 85.

16. Ralph A. Loomis and Glen T. Barton, Productivity of Agriculture:
United States, 1870-1958, USDA, Technical Bulletin No. 1238
(Washington, D.C., April, 1961), Tables 11 and 12, 57-58.

The net farm input series rises from an index of 85.7 in 1899
to 100 in 1929, while the nonfarm input series rises from 46.3
to 100 over the same period (1929 = 100).[17] Meiburg's data
suggest that farmers increased their reliance on nonfarm inputs
at a fairly steady rate over these thirty years; however, if
durable capital inputs had been included in the nonfarm input
series instead of the net farm input series, the proportionate
increase of nonfarm inputs relative to all production inputs
would probably be much greater in the 1920's than in the two
previous decades. The growth of durable capital in the 1920's is
disguised by the slight decline in the total stock of farm
capital in the United States during the decade. Almost all
of the latter decline was confined to the stock of work animals
and horses under three years of age. Stocks of machinery
and equipment rose at the same time, reflecting primarily the
shift after World War I from farm to nonfarm-produced power.[18]

Production inputs -- including farm and nonfarm produced --
are customarily divided into three broad classifications; land,

17. Charles O. Meiburg, "Nonfarm Inputs as a Source of Agricultural
Productivity," Food Research Institute Studies, Vol. III, No. 3
(November, 1962), Table 1, 219.

18. Alvin S. Tostlebe, Capital in Agriculture: Its Formation and
Financing Since 1870, NBER (Princeton, New Jersey, 1957), Table 9,
66-69.

labor and capital. Furthermore, all intermediate inputs not classified as land or labor are usually included in the capital category. As a percentage of all production inputs in American agriculture, land ranged from 18 to 20 per cent between 1900 and 1930. Labor fell from 57 per cent of all production inputs in 1900 to 46 per cent in 1930, while capital rose from 24 to 36 per cent.[19] The growing importance of capital relative to labor and land occurred steadily over these three decades; however, such data are so highly aggregated that important shifts in the components of each category are easily overlooked.

For example, historians frequently use data on the acreage of all land in United States farms to support the notion that "overexpansion" took place during World War I. Census data reveal that farm acreage increased slightly more than 17 per cent between 1900 and 1930; however, more than one-half of that increase, 8.8 per cent, occurred in the decade between 1910 and 1920.[20] Historians often attribute the 8.8 per cent increase

---

19. R.A. Loomis and G.T. Barton, Productivity of Agriculture, 50–51 and 58–59.

20. A.S. Tostlebe, Capital in Agriculture, Table 6, 50–51. In millions of acres, the figures are: 1900, 838.6; 1910, 878.8; 1920, 955.9; 1930, 986.8.

during the war decade to wartime expansion, and assume that these
new acres exacerbated the problem of "overproduction" in the
postwar decade. The material in previous chapters has cast
considerable doubt on the notion of "overproduction" in the
twenties; however, the idea of "overexpansion" in farm acreage durir
World War I merits additional attention.

It is not reasonable to attribute an increase in farm
acreage between two census dates to a war that began long after
the decade opened. Certainly some of the farm acreage added
between 1919 and 1920 must have been added before 1915, the
earliest point at which wartime demands could have influenced
production decisions. Unfortunately, expansion prior to 1915
is difficult to estimate since annual data on land in farms
exist only in a number of state agricultural bulletins.
One writer who compiled some of these state statistics suggested
that in states for which data are available, from one-fifth
to over one-half of all land added to farms in the ten years after
1910 was added before 1915.[21] Furthermore, only 35 million of the

21. Lloud P. Jorgenson, "Agricultural Expansion into the Semiarid
Lands of the West North Central States During the First World War,"
Agricultural History, Vol. 23, No. 1 (January, 1949), 32.

77 million acres added to farm land in the nation between 1910
and 1920 were added to harvested cropland. The rest, we must
infer, was added to pasture and woodland. Moreover, more than
30 million of the new acres added to cropland were added in
the Plains and Mountain regions.[22] High prices after 1915 were
only one factor that influenced settlement of those regions.
Many farmers were drawn to unsettled counties in those areas after
1910 because of dry-farming propaganda and a long period of
very favorable rainfall that did not end until 1917.[23] If
the 30 million acres opened in these western districts between
1910 and 1920 are disregarded, the total expansion of farm
land in the nation during that decade is reduced by about 40 per
cent. Much of the land opened in the Plains and Mountain regions
was of course seeded to wheat; however, the material in chapter IV
revealed that the postwar problems in these wheat districts
stemmed largely from drought and blight after 1917. Thus, the
notion of acreage "overexpansion" during the war leaves much
to be demonstrated.

Turning from land to another component of farm inputs,
labor, Table VI - 6 shows the regional pattern of labor inputs

22. Ibid., 33.
23. Mary Hargreaves, Dry Farming in the Northern Great Plains: 1900
1925 (Cambridge, Massachusetts, 1957), 473ff.

TABLE VI - 6

REGIONAL SHARES IN THE AGRICULTURAL LABOR FORCE OF THE
UNITED STATES, CENSUS YEARS, 1900-1930

	1900	1910	1920	1930
Northeast	11.8%	9.9%	9.1%	8.6%
Appalachian	16.7	16.4	15.6	15.5
Southeast	14.3	16.2	15.2	14.0
Lake States	8.2	7.2	8.0	8.0
Corn Belt	20.1	16.3	16.4	15.4
Delta States	10.7	11.9	11.1	11.8
Northern Plains	5.8	5.9	6.2	6.6
Southern Plains	7.9	10.3	10.3	11.0
Mountain	1.9	2.8	3.9	4.1
Pacific	2.5	3.1	4.1	5.0
U.S. (million persons)	11.3	12.4	10.7	10.5

Source: Calculated from data in Harvey Perloff, et. al., Regions,
Resources, and Economic Growth, 624-625.

in United States agriculture from 1900 to 1930. The relative
share of the nation's agricultural labor force in northern areas
east of the Mississippi and in the Appalachian region declined
steadily after 1900, while in the Southeast it declined only
after 1910. It is not surprising that all areas west of the
Mississippi registered gains throughout this period in their
respective shares of the agricultural labor force, with the
largest relative gains being in the Mountain and Pacific regions.
The westward shift of the agricultural labor force that continued
into the twentieth century reflected the availability of virgin
farm land, especially in the Mountain region and western sections
of the Northern Plains, as well as a relative absence of nonagricultural employment opportunities in these generally less
urbanized western areas.

A rough measure of the relative income producing power
of agricultural laborers in each region can be made by comparing
the data in Table VI - 6 with that shown in Table VI - 1.
In all southern regions, both east and west, the share of the
value of all farm products was considerably less than the share
of the agricultural labor force over the entire thirty year
period. This relationship was the opposite in all other regions,
indicating the wide gulf between the South and the rest of the

nation in farm practices, markets, and incomes. Although the
gulf may have narrowed somewhat over these thirty years, its
persistence is the more striking feature.

A possible clue to the disparity between north and south
in the value of farm output and the agricultural labor force
in offered in Table VI - 7 where real _physical_ capital (other
than land) per farm is shown for each region from 1900 to 1930.
In the six northern regions, where the share of value exceeds
the share of the labor force, the amount of capital per farm
is well above the national average at every date. In the four
southern regions, however, the reverse is true. This sort of
relationship was noted by Alvin Tostlebe in his study of capital
formation in agriculture when he compared per capita farm gross
income by region with real capital per person engaged in farming.

Tostlebe's data leave little doubt that there is a strong
positive relationship between capital and income in farming;
however, neither he nor anyone else has ever explained
satisfactorily why this is so.[24]

Economists generalize that greater use of capital inputs,
by permitting specialization and increasing the pace at which
a person can work, may raise the productivity of labor, without

24. A.S. Tostlebe, _Capital in Agriculture_, 94 and 97.

TABLE VI - 7

REAL PHYSICAL CAPITAL (OTHER THAN LAND) PER FARM,
BY REGIONS, CENSUS YEARS, 1900-1930
(hundreds of dollars)

	1900	1910	1920	1930
Northeast	$31.7	$34.1	$36.7	$41.5
Appalachian	12.0	13.5	15.3	14.7
Southeast	7.6	9.0	11.8	9.7
Lake States	23.9	32.7	39.1	42.2
Corn Belt	28.7	33.9	41.2	39.5
Delta States	9.3	9.0	10.4	8.6
Northern Plains	33.2	4.12	50.6	48.0
Southern Plains	18.0	17.0	21.4	17.4
Mountain	46.3	37.6	40.6	39.8
Pacific	31.1	32.3	34.2	33.3
United States	21.4	23.8	27.8	26.6

Source: Calculated from data in Alvin S. Tostlebe, Capital in
Agriculture, Table 6, 50-51 and Table 9, 66-69.

which an increase in real income is difficult to achieve.
Nevertheless, it does not seem likely that the income problem
of southern farmers revealed above could have improved substantially
if only their stock of physical capital had been augmented.
For one thing, the market situation in the South was different
from the North. Institutional arrangements connected with
securing capital often prevented timely adjustments from cotton
to livestock and dairy production, which in the long-run would
have yielded greater returns to most southern farmers.[25] Therefore,
the low income problem in southern agriculture, although intensified by shortages of capital inputs, was caused by a host of
other factors as well.

Coupled with the changes in farm inputs noted above,
went changes in the size of farm operations. In Table VI-8
the average acreage per farm and the percentage of all farmland
classified as "improved" is given by census year from 1900 to
1930. For the United States as a whole, the number of acres
per farm rose from 146 in 1900 to 157 in 1930, a gain of not
quite 9 per cent. Over the same period, the ratio of improved
farmland rose from just over 49 per cent of all land in farms to

25.   Harvey Perloff, et. al., <u>Regions, Resources, and Economic
Growth</u> (Baltimore, Maryland, 1960), 359-360. A.S. Tostlebe,
<u>Capital in Agriculture</u>, 98.

TABLE VI - 8

AVERAGE NUMBER OF ACRES PER FARM AND PERCENTAGE OF ALL FARMLAND
CLASSED AS "IMPROVED," BY REGIONS, CENSUS YEARS, 1900-1930

	1900	1910	1920	1930
Northeast				
acres per farm	96	96	99	102
improved percent	59	58	57	55
Appalachian				
acres per farm	102	91	84	80
improved percent	51	53	55	54
Southeast				
acres per farm	102	85	78	75
improved percent	41	45	49	53
Lake States				
acres per farm	121	125	127	130
improved percent	65	66	66	67
Corn Belt				
acres per farm	116	119	123	129
improved percent	79	81	80	78
Delta States				
acres per farm	89	76	71	60
improved percent	42	48	53	58
Northern Plains				
acres per farm	270	297	359	365
improved percent	61	66	69	68
Southern Plains				
acres per farm	323	232	232	227
improved percent	19	32	34	43
Mountain				
acres per farm	458	324	481	652
improved percent	18	27	26	30
Pacific				
acres per farm	335	270	240	231
improved percent	40	43	43	40
United states				
acres per farm	146	138	148	157
improved percent	49	54	53	54

Source: Calculated from data in Alvin S. Tostlebe, The Growth of
Physical Capital in Agriculture, 1870-1950, NBER, Occasional Paper
44, Table 10, 56.

to almost 53 per cent. Nevertheless, average farm size decreased over this period in the Pacific and southern regions -- increases in farm size were registered only in northern regions.

To some extent these regional differences are explained by shifts in commodity specialization and technological change. A major change in northern areas after 1900 was the shift to crop farming in the Plains and Mountain regions where large-scale units employing tractors and mechanized equipment were feasible. Current technological changes in eastern livestock and dairy areas were less likely to be reflected in the scale of land inputs. Therefore, increases in average farm size in the North between 1900 and 1930 were larger the farther west you moved. For example, acreage per farm grew from less than 7 per cent in the Northeast to more than 40 per cent in the Mountain region. On the other hand, a major change in southern areas after 1900 was the westward shift of cotton production which displaced a considerable amount of land - extensive livestock agriculture in the Delta and Southern Plains regions. Therefore, decreases in average farm size in the South from 1900 to 1930 were larger in western than in eastern areas.

In summary, this discussion of changes in the structure
of American farm inputs from 1900 to 1930 suggests that one
important transition which accelerated during the twenties was
the shift from farm to nonfarm produced inputs. A major factor in
that shift was the move to nonfarm produced power due primarily
to the adoption of internal combustion power sources. Along
with this, the growing capital intensity of American farms after
1900 spelled the need for more adequate farm finance. These
changes also spelled the need for more sophisticated farm management practices. Many farmers were faced with the decision of
whether or not to purchase tractors and, if they did invest
in such equipment, whether or not to expand the size of their
farms at the same time. Acreage expansion in turn raised the
issue of whether to buy more land outright or rent it. The
success with which farmers made these adjustments depended on
the available range of production alternatives, as well as each
farmer's skill and resources. In any event, the material discussed here leaves no doubt that countless American farmers had to
undertake adjustments of no small magnitude in the decade following
World War I.

Unfortunately, we do not have enough evidence to tell how

or by whom these adjustments were usually made. The data compiled

for this chapter can only suggest questions, not answers. Never-

theless, no historian writing about agriculture in the 1920's

has ever summarized or analyzed data even as highly aggregated

as that which is presented above. This is surprising, in view

of how much has been written about farm "crisis" in the twenties.

Output and input adjustments, when necessary, were undoubtedly

difficult to make, and data on farm bankruptcies suggest that

many farm operators were not successful in making these changes.

Yet, historians and economists have usually linked the evidence

for farm distress in the twenties to conditions arising out of

the wartime experience. No one has suggested that farm distress

may in many cases have resulted from inadequate adjustments to

changing demand and supply conditions.[26] Future research on

agriculture in the twenties must consider this more closely,

concentrating on the adjustments made by farmers in separate

commodity groups in local areas.

---

26. Another example of the type of adjustment problem discussed
here was that. faced by hog producers in the 1920's. The shift
in demand after the early 1900's from "lard-type" pork to
lean "bacon-type" pork had an impact on hog producers which has
never been studied and is never suggested as a possible cause
for stagnant incomes among producers who failed to adjust to the
change. See C.W. Towne and E.N. Wentworth, Pigs: From Cave to
Corn Belt (Norman, Oklahoma, 1950), chapter 19.

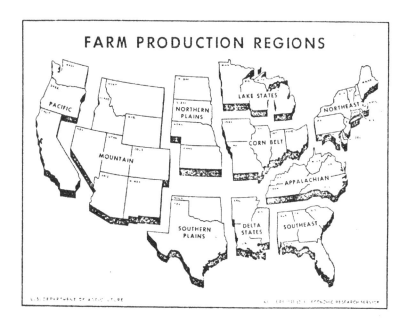

CHAPTER VII

AGRICULTURE, FARMERS AND ECONOMIC HISTORY:
FACTS AND ARTIFACTS

The main conclusion emerging from the discussion in
the six preceding chapters is that agricultural "depression,"
as that notion is applied to farm conditions in the 1920's,
represents a statistical artifact more than an economic fact.
Historians have accepted the idea that most American farmers
suffered economic depression from the end of 1920 until
after 1940, when wartime conditions brought recovery. For most
farmers, so it is said, conditions were not much different
in the twenties than they were later during the "great
depression." To support this assertion, scholars cite a wide
variety of quantitative data on farm incomes, prices and
financial distress. Nevertheless, the discussion in previous
chapters revealed a number of inadequacies in these statistics.
The main weakness found in most of the data was over-aggregation.

Too often the income, parity and other financial data used to describe the farmers' economic status include all farmers in the nation or in a region. The discussion of the wheat situation during the 1920's suggested that conclusions drawn from such highly aggregated data rarely explain particular farmers' grievances. Most historians would surely acknowledge that it is difficult, often impossible, to generalize from particular cases. Yet, agricultural historians often use generalized information to explain individual farmer's grievances. This tactic can produce misleading results, however, when the elements aggregated in a statistical average are as diverse as the collection of all farmers in the United States or a geographic region. Thus, agricultural historians and economists must focus more on results attained by homogeneous subsets of farmers within distinct commodity, soil and farm size classifications. This poses a challenge to scholars concerned with American agriculture after 1900, since much of the available data is already in a highly aggregated form.

Historians and economists have no doubt accepted the type of data that is criticized above because much of it

was prepared by an agency of the federal government, the U.S.
Department of Agriculture. The USDA's Bureau of Agricultural
Economics, especially during Henry C. Taylor's term as bureau
chief in the early twenties, supplied much of the information
that shaped and reinforced the traditional notion of post-
World War I agricultural "depression."[1] The credibitity and
prestige of "official" USDA material have no doubt been enhanced
by the excellent staff of statisticians and economists which
the Department has always employed. Authorities such as L.H. Bean
and Mordecai Ezekial are outstanding examples from the interwar
period. Nevertheless, in spite of the "official" stamp usually
assigned to government publications and data, historians should
not ignore the powerful support which the USDA and BAE gave to
George N. Peek in late 1923 and early 1924. Most of the

---

1. BAE data were frequently cited as basic source material
by agricultural experts writing in the 1920's and later.
Henry C. Taylor once boasted that BAE was the third largest
statistics-producing branch of the federal government, behind
the Bureau of the Census and the Weather Bureau. See The BAE
News, Vol. 9, No. 5 (October 16, 1923), filed at USDA, Agricultural History Branch.

statistical measures of agricultural "depression" that were

discussed in the previous chapters, including the "parity"

price index, as well as the usual interpretations attached

to these measures, were generated at the USDA.[2] Therefore,

if that agency was committed to a proposed policy framework

that was justified with data produced by the agency itself, it

is clear that interpretations and data that conflicted with

the "parity" notion of agricultural depression might have

been ignored by the Department.[3] Historians who rely exclusively

---

2. Major outlets for these data were the BAE's _The Agricultural Situation_, _Farm Real Estate Situation_ and _Crops and Markets_. The latter publication was carried-over from the former Bureau of Markets and Crop Estimates.

3. One economist recently noted that evaluations of policy which come from government agencies in general -- not just agricultural agencies -- are affected by a "...bias through exclusion from comparative analysis of policy systems that are alternatives to the existing ones. Policy discussion originating in a policy-making agency must be expected to define its terms of reference thus narrowly, for this limits the scope for criticism of the existing policy system by preventing its effective comparison with alternatives." See John M. Culbertson, _Macroeconomic Theory and Stabilization Policy_(New York, 1968), 418.

on USDA bulletins and data to depict agricultural conditions
in the 1920's have apparently overlooked this problem.[4]

Another possible reason why historians and economists
accept the conventional data on the agricultural "depression"
after World War I is that most of these data were developed
by social scientists who regarded fact-finding and statistics-

---

4. This problem was not overlooked, however, by Joseph S.
Davis, an economist with the Stanford Food Research Institute
who frequently criticized USDA and BAE statistics. He once
complained in a letter to O.C. Stine (USDA) that statistics
on farmer incomes shown in the July, 1927 Crops and Markets
were misleading. Here the gross returns of U.S. farmers
for each year of the 1920's were shown in index number fashion,
using 1919-1920 as a base. This of course showed seriously
depressed returns during the twenties as incomes in 1919
and 1920 were still at the wartime peak. After raising several
other points, Davis concluded:
"My reasons for expressing these criticisms at such length
lies (sic) partly in the fact that such tables seem to sum
up the gist of the position of farmers in comparison with certain
non-farming classes. They tend to be used very widely and
very loosely. If they are not only open to serious critisism on
such grounds as I have mentioned, but seriously misrepresent
the situation, as I believe they do (in the sense of misleading
many readers) the utmost care should be taken to perfect them
as fully as possible and in addition to couple with them as
effective qualifications as can be done. I believe that the
net effect of the errors (such as they are) I have mentioned is
to exaggerate the unfavorable position of the farmers. I
feel sure that the Department's policy leads it to guard
itself carefully against portraying the farm situation in too
favorable a light. I believe that it should be equally careful
to avoid portraying it in too unfavorable a light."
Letter from J.S. Davis to O.C. Stine (11/5/27) in U.S. National
Archives, Files of the BAE, Box 224, "Income 1."

gathering as important weapons of social reform.[5] The Pro-
gressive spirit of the early twentieth century emerged after
1920 in a guise that emphasized "scientific" statistical analysis
of social and economic problems as a prelude to reform. The
protest and reform posture assumed by most agricultural leaders
in the 1920's was an outstanding example of this trend.[6] Therefore, historians who write about agriculture in the twenties
have focused their research on the programs and leadership
of those reform and political pressure groups -- none have
scrutinized the economic analyses and quantitative data which
writers in the 1920's and 1930's used to explain agricultural
"depression."[7]

---

5. Henry F. May, "Shifting Perspectives on the 1920's,"
Mississippi Valley Historical Review (December, 1956), 408.

6. Donald L. Winters, "The Persistence of Progressivism: Henry
Cantwell Wallace and The Movement for Agricultural Economics,"
Agricultural History, Vol. XLI, No. 2 (April, 1967), 109-125.

7. There always are many plausible interpretations and explanations
for any social or economic problem. The fact that most scholars
since the 1920's have accepted one particular interpretation of
the "farm problem" in the United States suggests that the interpretation has never been adequately tested with meaningful
empirical evidence; however, economists must be judged more
at fault than historians for this oversight. John Culbertson,
referring to John M. Keynes's oft-quoted remark about "madmen
in authority ... distilling their frenzy from some academic
scribbler of a few years back," states that "economists and
political philosophers, in their generosity, provide ideas
to suit every taste and interest. Where one approach achieves
some predominance among economists, it may not be clear but
that this reflects some biasing factor, some subtle effective
'interest' of the academic profession -- or that the profession
is providing to people those ideas by which they currently
insist upon being 'enslaved.'"   J.M. Culbertson, _Macroeconomic_

There is no doubt that statistical data lend an air

of precision and finality to any arguments they are used to

support. Furthermore, statistics like the ones criticized in

the earlier chapters, which are aggregated at the national

level, appear to give a more complete picture than less highly

aggregated data would. This deceptive "comprehensive" quality

of macro-economic data is not limited of course to agricultural

statistics. William Leuchtenburg, in a bibliography on

many aspects of American history from 1914 to 1932, once noted

there were many gaps in the literature; however, after citing

many statistical studies of general economic conditions in the

1920's, he pronounced that the economic history of that era

had been studied "...to a fare-thee-well." Leuchtenburg dis-

counted the need for additional research on the economic history

of the twenties, saying that "thanks largely to the initiative

of the Brooking Institution and the National Bureau of Economic

Research, the student is faced chiefly with a problem of

Theory and Stabilization Policy, 520-531. With respect to
Culbertson's remarks, a very important article that was referred
to before is Vernon Carstensen, "An Historian Looks at the
Past Fifty Years of the Agricultural Economics Profession."

selection."[8] It is surprising that an outstanding historian

such as Professor Leuchtenburg could find definitive accounts

of American economic life in the "broad-brush" studies, often

based on data from federal government agencies, published by

Brookings and NBER. Nevertheless, Professor Leuchtenburg's

sentiments seem to be shared by the many scholars who have discussed

agricultural "depression" during the twenties in terms of the

ubiquitous parity ratio.

The parity ratio seems to give, at first glance, a

complete measure of changes in the welfare of the nation's

farmers. It is the ultimate quantification of the rural-

urban duality basic to the agrarian rhetoric of the nineteenth

century. Furthermore, when the parity index is compared with

numerous other series of national data, all of which reinforce

the conclusion that economic distress prevailed in the agricultural

sector in the 1920's, it would seem that the notion of farm

"depression" during the twenties was adequately supported. As

noted in chapter III, however, the parity ratio is an almost

---

8. William E. Leuchtenburg, The Perils of Prosperity, 1914-1932
(Chicago, 1958), 292-293.

meaningless measure of farmers' welfare.

Trends in commodity prices and input prices, from which the parity index is derived, usually give an ambiguous picture of the economic position of farm operators. For one thing, all farmers do not produce and market the same combination of commodities. Therefore, unless the prices of all commodities move in unison and unless production yields among farm operators are identical, trends in _average_ commodity prices reflect little or nothing about the gross receipts of individual farmers. Likewise, all farmers do not employ the same combination of inputs on farms of the same size. That being the case, trends in average input prices tell us nothing about the unit costs of individual farmers. Finally, if all other things were equal, the parity ratio would still fail as an index of farmers' economic welfare because it ignores changes in input productivity.

The parity ratio stands, therefore, as the prime example of a statistical artifact from which countless scholars have drawn generalizations about economic conditions in American agriculture after World War I. As a result, these scholars

have failed to note that prices are affected by both supply
and demand forces, and that it is important to know which
forces were at work before concluding that "unfavorable"
price trends caused widespread economic distress among farmers
in the 1920's.

If firms are producing in an industry near a position
of long-run equilibrium and price falls permanently because
of decreased demand, we would expect many firms to suffer
financial distress and even bankruptcy. Presumably this happened
in American agriculture after World War I, if one accepts
the "parity" notion of overproduction, low prices and agricultural
depression in the 1920's. The discussion in chapter VI
revealed many cases, however, where decreased demand for one
commodity, such as hay, was offset by increased demand for
another commodity, such as milk, that used the former product
as an input. Where this happened, falling product prices would
not necessarily have led to farm failures if individual
farmers made the necessary adjustments. As noted before, we
cannot tell from available sources how much farm "distress"

in the twenties resulted from failure or inability to make these adjustments. Future research in farm records, where available, may shed some light on the issue.

Prices can also be affected by supply conditions as well as demand conditions. If market price falls because firms have increased industry supply by adopting new cost-saving inputs, then our predictions about firm failures would depend on things such as distribution of firm size in the industry and the capital outlay required to adopt the new inputs. A case in point during the 1920's was the adoption of tractor-powered combines on western wheat farms, discussed in chapter IV. The combine helped reduce unit costs on farms in the Southern Plains; however, it probably induced an increase in industry supply as well. If so, hard winter wheat prices would have been affected, although hard spring wheat prices would probably have been affected less since the two varieties were priced on more or less different bases at that time. In any case, downward pressure on hard winter wheat prices would not necessarily have been cause for economic distress among wheat farmers who could adopt the new machinery. Some producers no doubt failed because they were unable or not willing to adjust their inputs; however, historians have been silent on the matter, preferring instead to blame most farm failures during the twenties on low prices and overproduction.

Thus, the effects of a change in commodity price can vary, depending on whether the change was caused primarily be a shift in industry supply or a shift in market demand. Supply and demand forces have never been clearly specified by scholars who write about agricultural "depression" during the 1920's. Even if supply forces and demand forces during the twenties were accurately specified, however, it would be almost impossible to pinpoint the effects of the price changes induced by such forces without additional data on specific farm operations. The plethora of price and production data now available are too highly aggregated for such analysis. We need farm operating data from the twenties that are classified by distinct commodity and soil characteristics, data which perhaps will be available only through samples drawn from manuscript census records.

Sample data from the census may also help us estimate the extent of farm distress caused during the 1920's by optimistic capitalization of high commodity prices between 1917 and 1920.

There is no doubt that net incomes of many, if not most farmers rose to exceptionally high levels during the war. From an accounting standpoint, gross receipts increased far more than production costs, reflecting a proportionately larger rise in commodity prices than in average total costs. Even from an economic standpoint, many farm operators probably earned "profit" during the wartime period. In other words, market prices were so high that in many cases farmers' total incomes exceeded what might have been earned in the best alternative use of their resources. Some farmers, mentioned above in chapter V, capitalized this "profit" into higher fixed costs when they purchased more farm land or mortgaged their existing farm at values attained in 1919 or early 1920. Other farmers entered the 1920's with essentially the same fixed cost base they had when the war began. Perhaps much of the economic distress suffered by farmers after 1920 was confined to farm operators in the former category. For those farmers, the decline in variable costs after 1920, although as steep as the fall in commodity prices in most cases, was not enough to compensate for the enormous rise in fixed costs that occurred during the postwar boom. It is not appropriate to say that this group suffered from low prices because of over-production during the twenties. Yet, historians writing about agricultural

"depression" say that when they fail to distinguish between

those farmers who did and those who did not capitalize the

high earnings of the war years.

If historians and economists paid closer attention to

the diversity of problems that American farmers faced after World

War I, and concentrated less on aggregate price, production

and financial data, it is probable they would discard the concept

of "overproduction." This concept, never clearly defined by

those who use it, presumes that farm commodity prices are rarely

high enough to insure farm producers a return on their time and

investment that is as high as they could earn in alternative

nonfarm employment. USDA data comparing farm and nonfarm incomes

per capita have always supported this presumption. Furthermore,

the willingness of many farmers to remain on the farm when they

supposedly could earn more elsewhere is usually explained by

psychic advantages of farm life. Nevertheless, the discussion

in chapter II revealed that the gap between farm and nonfarm

earnings is not as glaring as USDA data show, after adjustments

that make the two series of data comparable are considered.

Additional suspicion was cast on the notion that "over-production" caused low farm incomes in the twenties (and later) when we noted that national farm income averages in the United States are severely depressed by the _very_ low earnings of many small farmers, especially in the South. These "farmers" impute a low return to their labor and consequently are less willing to leave the land than if their assessment of foregone opportunities were more realistic.[9] But this is a general economic and social problem rather than an agricultural problem. The low incomes of these rural poor did not reflect low prices and overproduction in the 1920's any more than they do today. Policy-makers today are usually aware of these ambiguities in national farm income statistics; however, historians writing about agricultural"depression" in the 1920's should stop citing such highly aggregated data as proof that low prices and over-production caused most farmers' grievances after World War I.

---

9. The unique problems of rural poor in the American South are discussed at many places in John McKinney and Edgar T. Thompson, _The South in Continuity and Change_ (Durham, North Carolina, 1965) and Thomas R. Ford, ed., _The Southern Appalachian Region_ (Lexington, Kentucky, 1962).

It is apparent from all that has been said here, that
we understand very little about the roots of farm protest in
the 1920's. Perhaps much of the protest was registered on
behalf of special interest groups, such as rural banks and
cooperative marketing organizations, whose problems often
stemmed from a myriad of conditions not related to those
faced by farmers themselves. On the other hand, many farmers
as such _did_ experience difficulties in the twenties, although
the previous chapters suggest that the relative importance of
various economic causes for such problems remains to be demonstrated.

Non-economic causes of farm protest in the 1920's
have not been discussed here; however, there surely must have
been several. Some agricultural economists, writing in the
late 1920's, suggested that the psychological depression of farmers was more real than the widely heralded financial depression.[10]

---

10. Joseph S. Davis, "America's Agricultural Position and
Policy" (1927), in On Agricultural Policy: 1926-1938 (Stanford,
California, 1939). J. D. Black, Agricultural Reform in the
United States, 23.

Although farm incomes on the whole were rising, perhaps as
rapidly as nonfarm incomes, many farmers could not use their
income to acquire the many consumer durables that were making
the urban resident's life more enjoyable. Most of these
new goods required electricity to operate, but few farmers
could enjoy the benefits of electric power in this period.
Also many public services such as hospitals, schools, and paved
roads were available to farmers only at great cost, if at all.
The farmer's relatively greater isolation precluded putting
his income to the same uses as an urban resident with the same
income. Today this problem is much less apparent, but in the
1920's it undoubtedly was a source of considerable frustration
for farm residents.

Finally, many farmers in the 1920's had to accept
stability as a way of life for the first time.[11] Farmers
always had been faced with the disruptive influence of
changing technology in production and marketing as well as
changing consumer demands. Before about 1910, farmers who did not
adapt to these changes could move on to new areas, however,
where lower land values and more favorable production conditions

11. J.C. Malin, The Grassland of North America, 306ff. and
passim.

made it possible to operate profitably with the old methods. In the process, the farmer who moved continuously from one frontier to the next often found that economic gain from land appreciation was an important form of income.[12] By 1920, opportunities for mobility were rapidly diminishing, and falling land values after 1920 prevented many who might have moved from doing so. These conditions were further aggravated by the postwar technollogical and market demand changes that made the situation especially hard for farmers unable to adapt. If such farmers contributed to the foreclosure and bankruptcy statistics, then the cause for their distress had existed in agriculture long before 1920, but its presence had always been masked by the movement of marginal farmers to new areas.

We can only speculate about the relative importance of the many causes for distress which clearly did exist. We know all too little about which farmers faced economic hardship in the twenties and why. Likewise, we know little about farmers who adapted to change successfully and profited in the decade. Historians have all but ignored the latter

12. See above, page 134.

group, for whom the story told by the exponents of the "parity" concept had little relevance. More could be learned, not only about conditions in the 1920's but in other periods as well, be devoting more research effort to the economic affairs of farmers as such and less to the statistical artifact known as "agriculture."

BIBLIOGRAPHY

Harold U. Faulkner's observation that "no full-length
history of American agriculture has been attempted covering
the period since 1897," is as valid today as it was almost twenty
years ago when he first made it. Fortunately there are several
excellent bibliographies and indexes to the literature in the field,
of which the following three are outstanding examples: The American
Economic Association, Index of Economic Journals: Everett
E. Edwards, A Bibliography of the History of Agriculture in the
United States; Agricultural Index. Equally notable are the
voluminous bibliographical files maintained by Mrs. Helen
Edwards at the U.S. Department of Agriculture, Agricultural
History Branch. The research for this thesis included an examin-
ation of the above sources as well as a review of all volumes of
Agricultural History, Journal of Farm Economics, and U. S.
Department of Agriculture Yearbooks published since 1900.
In addition, several yearbooks and bulletins of various state
agricultural departments and experiment stations were reviewed.
The following list includes all items cited in the thesis as
well as several important sources not specifically mentioned herein
which were important to the research. Two abbreviations used
below are USDA (U.S. Department of Agriculture) and NBER
(National Bureau of Economic Research).

MANUSCRIPT COLLECTIONS AND UNPUBLISHED
GOVERNMENT STATISTICS

Papers of the Bureau of Agricultural Economics. Record Group
83, U.S. National Archives. Washington, D.C.

The Henry C. Taylor Papers. State Historical Society of Wisconsin.
Madison, Wisconsin.

Worksheets showing farm distress transfer rates, farm land
values, and farm land rents by crop reporting districts in
the United States, various dates from 1912 to 1933. USDA,
Economic Research Service, Farm Production Economics
Division. Washington, D.C. These data were used to prepare
editions of The Farm Real Estate Situation.

Bank suspension statistics by county for the United States,
1920-1933. Works Progress Administration, Federal Deposit
Insurance Corporation. On indefinite loan to the Graduate
Program in Economic History. Madison, Wisconsin.

GOVERNMENT PUBLICATIONS

Baker, O.E. A Graphic Summary of the Number, Size, and Type of
Farm, and Value of Products. USDA. Misc. Publication No. 266.
Washington, D.C., October, 1937.

Board of Governors of the Federal Reserve System. Federal
Reserve Bulletin, Vol. 23: No. 9 (September, 1937), 878-906,
and No. 12 (December, 1937), 1204-1224.

Chambers, Clyde R. Relation of Land Income to Land Value. USDA.
Bulletin No. 1224. Washington, D.C., June, 1924.

Church, L.M. Farm Motor Truck Operation in the New England
and Central Atlantic States. USDA. Department Bulletin No.
1254. Washington, D.C., September, 1924.

GOVERNMENT PUBLICATIONS (contd.)

Cooper, Martin R., Glen T. Barton and Albert P. Brodell. Progress of Farm Mechanization. USDA. Misc. Publication 630. Washington D.C., October, 1947.

Gray, L.C. and O.G. Lloyd. Farmland Values in Iowa. USDA. Bulletin No. 874. Washington, D.C., 1920.

Hecht, Reuben W. and Glen T. Barton. Gains in Productivity of Farm Labor. USDA. Technical Bulletin No. 1020. Washington, D.C., December, 1950.

Hopkins, John A. Changing Technology and Employment in Agriculture. USDA. Washington, D.C., May, 1941.

Horton, D.C., H.C. Larsen and N.J. Wall. Farm-Mortgage Credit Facilities in the United States. USDA. Misc. Publication No. 478. Washington, D.C., 1942.

Iden, George. "Farmland Values Reexplored." USDA. Agricultural Economics Research, Vol. XVI, No. 2 (April, 1964), 41-50.

Loomis, Ralph and Glen T. Barton. Productivity of Agriculture; United States, 1870-1958. USDA. Technical Bulletin No. 1238. Washington, D.C., April, 1961.

Meyers, Albert L. "Agriculture and the National Economy." Temporary National Economic Committee, Investigation of Concentration of Economic Power, Monograph No. 23. Washington, D.C., 1940.

Miller, Herman P. Income Distribution in the United States. U.S. Department of Commerce, Bureau of the Census. Washington, D.C., 1966.

Swanson, Joseph A. "Economic Growth and the Theory of Agricultural Revolution." USDA. Agricultural Economics Research, Vol. XVI, No. 2 (April, 1964), 51-56.

Tolley, H.R. and L.M. Church. Motor Trucks on Corn Belt Farms. USDA. Farmers' Bulletin No. 1314. Washington, D.C., March, 1923.

GOVERNMENT PUBLICATIONS (contd.)

U.S. Congress, Joint Commission of Agricultural Inquiry. Report of the Joint Commission of Agricultural Inquiry, Part I. House Report No. 408, 67th Congress, 1st Session. Washington, D.C., 1921.

U.S. Department of Agriculture. Agriculture Yearbook. Washington, D.C., 1919-1929.

U.S. Department of Agriculture. Crops and Markets, Monthly Supplement. Washington, D.C., July, 1925-July, 1927.

U.S. Department of Agriculture. Major Statistical Series of the USDA, Agriculture Handbook No. 118, 10 volumes. Washington, D.C., 1957.

U.S. Department of Agriculture. "Wheat: Acreage, Yield and Production by States, 1866-1943." Statistical Bulletin No. 158 Washington, D.C., February, 1955.

U.S. Department of Agriculture. Yearbook of Agriculture: 1931. Washington, D.C., 1931.

U.S. Department of Agriculture, Bureau of Agricultural Economics. Income Parity For Agriculture, mimeograph in 6 parts. Washington, D.C., 1938-1945.

U.S. Department of Commerce, Bureau of the Census. Historical Statistics of the United States: Colonial Times to 1957. Washington. D.C., 1960.

U.S. Department of Commerce. Census Office. Twelfth Census of the U.S., 1900. Agriculture. Washington, D.C., 1902.

U.S. Department of Commerce.Bureau of the Census. Fourteenth Census of the U.S., 1920. Agriculture. Washington, D.C.,

U.S. Department of Commerce. Bureau of the Census. U.S. Census of Agriculture: 1950. Washington, D.C., 1953.

U.S. Works Progress Administration. Changes in Technology and Labor Requirements in Crop Production: Wheat and Oats. National Research Project Report No. A-10. Philadelphia, April, 1939.

233.

GOVERNMENT PUBLICATIONS (contd.)

Wall, Norman J. and Lawrence A. Jones. "Short-Term Agricultural
Loans of Commercial Banks, 1910-1945." USDA.Agricultural
Finance Review, Vol. 8 (November, 1945), 2-5.

Wallace, Henry C. "The Wheat Situation." USDA. Agriculture Yearbook:
1923, 95-150. Washington, D.C., 1924.

Wickens, David L. Farmer Bankruptcies, 1898-1935. USDA.
Circular No. 414. Washington, D.C., September, 1936.

Wiecking, E.H. The Farm Real Estate Situation, 1929-30. USDA.
Circular No. 150. Washington, D.C., November, 1930.

BOOKS, PAMPHLETS AND SELECTIONS FROM BOOKS

The American Academy of Political and Social Sciences. The Annals.
"The Agricultural Situation in the United States." Vol. CXVII.
Philadelphia, 1925. "Farm Relief." Vol. CXLII. Philadelphia,
1929.

American Society of Agricultural Engineers. Present Status of
"Combine" Harvesting. St. Joseph, Michigan, March, 1928.

Barger, Harold and Hans H. Landsberg. American Agriculture, 1899-
1939: A Study of Output, Employment and Productivity. NBER.
New York, 1942.

Benedict, Murray R. Farm Policies of the United States: 1790-1950.
New York, 1953.

Black, John D. Agricultural Reform in the United States.
New York, 1929.

Black, John. D. Parity, Parity, Parity. Cambridge, Massachusetts,
1942.

BOOKS, PAMPHLETS AND SELECTIONS FROM BOOKS
(contd.)

Bogue, Allan G. From Prairie to Corn Belt. Chicago, 1963.

Bolino, August C. The Development of the American Economy. Columbus
Ohio, 1966.

Boyle, James E. Farm Relief: A Brief on the NcNary-Haugen Plan.
New York, 1928.

Bremer, C.D. American Bank Failures. New York, 1935.

Business Men's Commission on Agriculture. The Condition of
Agriculture in the United States and Measures for its
Improvement. New York and Washington, D.C., 1927.

Collins, Charles W. Rural Banking Reform. New York, 1931.

Culbertson, John M. Macroeconomic Theory and Stabilization
Policy. New York, 1968.

Davis, Joseph S. On Agricultural Policy: 1926-1938. Stanford,
California, 1939.

Easterlin, Richard A. "State Income Estimates." Simon Kuznets
and Dorothy Swaine Thomas, Population Redistribution and
Economic Growth: United States, 1870-1950. Vol. I. Philadelphia,
1957.

Everitt, J.A. The Third Power. Indianapolis, Indiana, 1903.

Faulkner, Harold U. American Economic History. New York, 1943.

Fite, Gilbert C. George N. Peek and the Fight for Farm Parity.
Norman, Oklahoma, 1954.

Ford, Thomas R., ed. The Southern Appalachian Region. Lexington,
Kentucky, 1962.

Gates, Paul W. The Farmer's Age: Agriculture, 1815-1860. New York,
1960.

BOOKS, PAMPHLETS AND SELECTIONS FROM BOOKS
(contd.)

Genung, A.B. The Agricultural Depression Following World War I
and its Political Consequences: An Account of the Deflation
Episode, 1921-1934. Ithaca, New York, 1954.

Goodrich, Carter. Migration and Economic Opportunity. Philadelphia,
1936.

Greenleaf, William, ed. American Economic Development Since 1860.
New York, 1968.

Hacker, Louis M. The Shaping of the American Tradition. Text
edition. New York, 1947.

Halcrow, Harold G. Agricultural Policy of the United States.
New York, 1953.

Hargreaves, Mary. Dry Farming in the Northern Great Plains:
1900-1925. Cambridge, Massachusetts, 1957.

Heady, Earl O. and Luther G. Tweeten. Resource Demand and
Structure of the Agricultural Industry. Ames, Iowa, 1963.

Heady, Earl O., et. al. Roots of the Farm Problem. Ames,
Iowa, 1965.

Hibbard, Benjamin H. Effects of the Great War Upon Agriculture
in the United States and Great Britain. Carnegie Endowment for
International Peace. New York, 1919.

Johnson, D. Gale. "Economics of Agriculture." Bernard F. Haley,
A Survey of Contemporary Economics, Vol. II, 223-260. Homewood,
Illinois, 1952.

Jones, Lawrence A. and David Durand. Mortgage Lending Experience
in Agriculture. NBER. Princeton, New Jersey, 1954.

BOOKS, PAMPHLETS AND SELECTIONS FROM BOOKS
(contd.)

Kemmerer, Donald L. and C. Clyde Jones. American Economic History.
New York, 1959.

Kendrick, John W. Productivity Trends in the United States. NBER.
Princeton, New Jersey, 1961.

King, Willford I. The National Income and its Purchasing
Power. NBER. New York, 1930.

King, Willford I. Wealth and Income of the People of the United
States. New York, 1922.

Kuznets, Simon. National Product in Wartime. NBER. New York, 1945.

Lave, Lester B. Technological Change: Its Conception and
Measurement. Englewood Cliffs, New Jersey, 1966.

Leutchtenburg, William E. The Perils of Prosperity, 1914-1932.
Chicago, 1958.

Link, Arthur S. American Epoch: A History of the United States
Since the 1890's. New York, 1955.

Malin, James C. The Grassland of North America: Prolegomena to its
History. Lawrence, Kansas, 1947.

Malin, James C. Winter Wheat in the Golden Belt of Kansas.
Lawrence, Kansas, 1944.

McKinney, John and Edgar T. Thompson. The South in Continuity and
Change. Durham, North Carolina, 1965.

Mills, Frederick C. Economic Tendencies in the United States: Aspect
of Pre-War and Post-War Changes. NBER. New York, 1932.

Nourse, E.G. "Agriculture." Committee on Recent Economic Changes,
Recent Economic Changes in the United States, chapter VIII.
New York, 1929.

BOOKS, PAMPHLETS AND SELECTIONS FROM BOOKS
(contd.)

Nourse, Edwin G. American Agriculture and the European Market.
New York, 1924.

Panschar, William G. Baking in America. Vol. I. Evanston,
Illinois, 1956.

Parker, William and Judith Klein, "Productivity Growth in Grain
Production in the United States, 1840-1860 and 1900-1910."
NBER Output, Employment and Productivity in the United
States after 1800, 523-582. Princeton, New Jersey, 1966.

Peek, George N. and Hugh S. Johnson. Equality for Agriculture.
Moline, Illinois, 1922.

Perkins, Van L. Crisis in Agriculture: The Agricultural Adjustment
Administration and the New Deal, 1933. Berkeley, California,
1969.

Perloff, Harvey; E. Dunn, E.E. Lampard and R. Muth. Regions,
Resources, and Economic Growth. Baltimore, Maryland, 1960.

Ross, Earle D. "Agriculture in an Industrial Economy."
Harold F. Williamson, ed., The Growth of the American Economy,
687-698. Englewood Cliffs, New Jersey, 1951.

Sakolski, Aaron M. The Great American Land Bubble. New York, 1932.

Saloutos, Theodore and John D. Hicks. Agricultural Discontent
in the Middle West, 1900-1939. Madison, Wisconsin, 1951.

Samuelson, Paul A. and Everett E. Hagen. After the War: 1918-1920.
National Resrouces Planning Board. Washington, D.C., 1943.

Schultz, Theodore. Agriculture in an Unstable Economy. New York, 1945.

Seligman, Edwin R.A. The Economics of Farm Relief. New York, 1929.

Shideler, James H. Farm Crisis: 1919-1923. Berkeley and Los
Angeles, California, 1957.

Silberling, Norman J. The Dynamics of Business. New York, 1943.

BOOKS, PAMPHLETS AND SELECTIONS FROM BOOKS
(contd.)

Singer, Eugene M. Antitrust Economics: Selected Legal Cases
and Economic Models. Englewood Cliffs, New Jersey, 1968.

Soule, George. Prosperity Decade, From War to Depression: 1917-
1929. New York, 1947.

Stevenson, Russell A., ed. A Type Study of American Banking:
Non-Metropolitan Banks in Minnesota. Minneapolis, November, 1934

Taylor, Henry C. and Anne Dewees. The Story of Agricultural
Economics in the United States: 1840-1932. Ames, Iowa. 1952.

Tostlebe, Alvin S. Capital in Agriculture: Its Formation and
Financing Since 1870. NBER. Princeton, New Jersey, 1957.

Tostlebe, Alvin S. The Growth of Physical Capital in Agriculture.
NBER. Occasional Paper 44. New York, 1954.

Towne, C.W. and E.N. Wentworth. Pigs: From Cave to Corn Belt.
Norman, Oklahoma, 1950.

Wallace, Henry A. Democracy Reborn. ed. by Russell Lord. New
York, 1944.

Wallace, Henry C. Our Debt and Duty to the Farmer. New York, 1925.

Warren, George F. and Frank A. Pearson. The Agricultural Situation:
Economic Effects of Fluctuating Prices. New York, 1924.

Weiss, Leonard W. Economics and American Industry. New York, 1961.

Wiley, Clarence A. Agriculture and the Business Cycle Since 1920:
A Study in the Post-War Disparity of Prices. Madison, Wisconsin,
1930.

Wright, Chester W. Economic History of the United States. New
York, 1949.

ARTICLES FROM PERIODICALS

Anderson, W.J. "The Basis of Economic Policy for Canadian Agriculture."
Canadian Journal of Agricultural Economics, Vol. XI, No. 2 (April,
1963), 19-28.

Black, John D. "Income of Farmers - Discussion." American
Economic Review, Supplement, Vol. XIII, No. 1 (March, 1923),
181-184.

Bowman, John D. "An Economic Analysis of Midwestern Farm Land
Values and Farm Land Income, 1860-1900." Yale Economic
Essays, Vol. 5, No. 2 (Fall, 1965), 317-352.

Carstensen, Vernon "An Historian Looks at the Past Fifty
Years of the Agricultural Economics Profession." Journal
of Farm Economics, Vol. 42, No. 5 (December, 1960), 994-1006.

Coulter, John Lee. "The Wheat Crisis." The Quarterly Journal
of the University of North Dakota, Vol. 14, No. 1 (November,
1923), 3-26.

Egerer, Gerald. "The Political Economy of British Wheat, 1920-1960."
Agricultural History, Vol. XL, No. 4 (October, 1966), 295-310.

Englund, Eric. "The Bank's Part in the Farmer's Trouble."
Nation's Business, Vol. XIV, No. 11 (October, 1926),
13ff.

Food Research Institute. "Disposition of American Wheat Since
1896." Wheat Studies of the Food Research Institute, Vol. IV,
No. 14. (February, 1928).

Food Research Institite. "The World Wheat Situation, 1928-29."
Wheat Studies of the Food Research Institute, Vol. VI, No. 2
(December, 1929).

Garlock, Fred L. "Bank Failures in Iowa." Journal of Land and Public
Utility Economics, Vol. II, No. 1 (January, 1926), 48-66.

Grimes, W.E. "Some Phases of the Hard Winter Wheat Grower's
Problem in Readjustment." Journal of Farm Economics,
Vol. VII, No. 2 (April, 1925), 196-219.

ARTICLES FROM PERIODICALS
(contd.)

Grimes, W.E. "Trends in the Agriculture of the Hard Winter Wheat
    Belt." The Journal of Land and Public Utility Economics, Vol. I*
    No. 4 (November, 1928), 347-354.

Jorgenson, Lloyd, P. "Agricultural Expansion Into the Semiarid
    Lands of the West North Central States During the First World
    War." Agricultural History, Vol. 23, No. 1 (January, 1949),
    30-33.

Lloyd, O.G. "Studies of Land Values in Iowa." Journal of Farm
    Economics, Vol. 2, No. 3 (July, 1920), 136-140.

Malin, James C. "Mobility and History: Reflections on the
    Agricultural Policies of the United States in Relation to
    a Mechanized World." Agricultural History, Vol. 17,
    No. 4 (October, 1943), 177-191.

May, Henry F. "Shifting Perspectives on the 1920's." Mississippi
    Valley Historical Review, Vol. XLIII, No. 3 (December, 1956),
    405-427.

McDonald, Stephen L. "Farm Outmigration as an Integrative
    Adjustment to Economic Change." Social Forces, Vol. 34,
    No. 2 (December, 1955), 119-128.

Meiburg, Charles O. "Nonfarm Inputs as a Source of Agricultural
    Productivity." Food Research Institute Studies, Vol. III,
    No. 3 (November, 1962), 217-221.

Meiburg, Charles O. and Karl Brandt. "Agricultural Productivity
    in the United States: 1870-1960." Food Research Institute Studie
    Vol. III, No. 2 (May, 1962), 63-85.

Murray, William G. "Iowa Land Values: 1803-1967." The Palimpsest,
    Vol. XLVII, No. 10 (October, 1967), 441-504.

Norton, L.J. "The Land Market and Farm Mortgage Debts, 1917-1921."
    Journal of Farm Economics, Vol. 24, No. 1 (February, 1942),
    168-177.

O'Brien, Harry R. "Iowa's Abandoned Farms: A Study of How the
    Inflated, Speculative Prices of Land Affect the State."
    The Country Gentleman (June 18, 1921), 8ff.

ARTICLES FROM PERIODICALS
(contd.)

Owen, Wyn F. "The Double Developmental Squeeze on Agriculture."
American Economic Review, Vol. LVI, No. 1 (March, 1966),
43-70.

Peck, H.W. "The Influence of Agricultural Machinery and the
Automobile on Farming Operations." The Quarterly Journal
of Economics, Vol. XLI (May, 1927), 534-544.

Peterson, Arthur G. "Governmental Policy Relating to Farm
Machinery in World War I," Agricultural History, Vol. 17,
No. 1 (January, 1943), 21-40.

Ross, Earle D. "Retardation in Farm Technology Before the Power
Age." Agricultural History, Vol. 30, No. 1 (January, 1956),
11-18.

Shideler, James H. "The Development of the Parity Price Formula
for Agriculture." Agricultural History, Vol.XXVII, No. 3
(July, 1953), 77-84.

Studensky, G.A. "The Agricultural Depression and the Technical
Revolution in Farming." Journal of Farm Economics, Vol. 12,
No. 4 (October, 1930), 552-572.

Taylor, Alonzo E. "The Dispensability of a Wheat Surplus in
the United States." Wheat Studies of the Food Research
Institute, Vol. I, No. 4 (March, 1925), 121-142.

Tolley, H.R. "The Farm Power Problem." Journal of Farm Economics,
Vol. III, No. 2 (April, 1921), 91-99.

Warren, George F. "Some After-The-War Problems in Agriculture."
Journal of Farm Economics, Vol. I, No. 1 (June, 1919), 12-23.

Willard, Rex E. "Comments." Journal of Farm Economics, Vol. VII,
No. 2 (April, 1925), 220-221.

Winters, Donald L. "The Persistence of Progressivism: Henry
Cantwell Wallace and the Movement for Agricultural Economics."

ARTICLES FROM PERIODICALS
(contd.)

Agricultural History, Vol. XLI, No. 2 (April, 1967), 109–120.

THESES AND UNPUBLISHED PAPERS

Dost, Jeanne E. An Interregional Analysis of the Three Major Wheat Producing Regions of the United States. Ph.D. thesis. Radcliffe College, 1958.

Keehn, Richard H. "Agricultural Sector Terms of Trade for Four Midwestern States: 1870–1900," mimeograph. Madison, Wisconsin, 1966.

Mowry, G.E. The Decline of Agriculture, 1920–1924. M.A. thesis, University of Wisconsin, 1934.

Sjo, John Bernard. Technology: Its Effect on the Wheat Industry, Ph.D. thesis. Michigan State University, 1960.

Warburton, Clark. Deposit Insurance in Eight States: During the Period 1908–1930, manuscript. Washington, D.C., FDIC library, 1959.

Warburton, Clark, "Eleven Years of Research at the Federal Deposit Insurance Corporation," manuscript. Washington, D.C., FDIC library (n.d.).

Winters, Donald L. Henry Cantwell Wallace and the Farm Crisis of the Early Twenties. Ph.D. thesis. University of Wisconsin, 1966